# THE MOON IS NEW

## TIME COMES IN WITH A MINUS SIGN

John Lowry Dobson

Berbeo Publishing
2008

THE MOON IS NEW
TIME COMES IN WITH A MINUS SIGN



Published by
Berbeo Publishing
PO Box 66874
Los Angeles, California 90066

www.TheMoonIsNew.com

Cover Illustration: NGC 4565 with no foreground stars by MaryAnn Arrien,
used by permission.
Layout and arrangement by Eileen Berbeo
First copyright © 2004

ISBN # 978-0-9816952-0-4

Library of Congress Control Number: 2008926007

# PREFACE

We don't have two worlds, one for the mystics and one for the scientists. There's only one of it. And if the mystics are right in their descriptions, and if the scientists are right in *theirs*, we need only a translator and a dictionary of both languages. Fortunately for me, I have lived and worked in both camps and I now know both languages, and this story is my translation. But there's still a severe problem: How to write the science so the mystics can read it, and how to write the mysticism so the scientists can read it? Probably neither of these is possible at the present time, but we try.

But there's another problem: Not all the scientists have all the science straight, and not all the mystics have all the mysticism straight. So the girl in the story, who knows her physics, straightens out our mysticism through her physics, and some of ours. You'll see.

The task of putting science and religion together was assigned to two of us nearly fifty years ago while I was living in the monastery of the Vedanta Society of Northern California in San Francisco under Swami Ashokananda. He was an outstanding follower of Swami Vivekananda who represented Hinduism at the Parliament of Religions in Chicago in 1893 and became their most popular speaker. It was he who suggested to Nikola Tesla that what we call matter could be reduced to potential energy. That notion went from Tesla through Mileva Maric, Einstein's first wife, to the equations of relativity theory in 1905.

Swami Vivekananda wanted science and religion put together, but European science wasn't ready in his day, so Swami Ashokananda assigned the job to Michael Fell and me in 1955.

Swami Vivekananda, often called Swamiji, was a disciple of an Indian saint, Sri Ramakrishna, whom the girl in the story referred to as the Old Man in J.D.'s shrine. He lived in Bengal, India from 1836 till 1886 and practiced many sorts of spiritual practices and saw that they all lead to the same goal. He left a band of monastic disciples in the hands of Swamiji when he left. One of that band, Swami Premananda, is referred to in the story. (The word swami is the Sanskrit word for monk.)

Several equations will be referred to in the story, first, Einstein's famous equation which simply says that what we call mass (matter) is just potential energy.

Then there is Einstein's 1905 geometry which puts time in as anti-space; so that to get the *objective space-time separation* between two events we have to subtract the time separation from the space separation. Just as up is the opposite of down, just so time is the opposite of space. And if a light beam, or a gravitational message, can get from one event to another then the space and time separations between those two events are equal and the total space-time separation is zero.

And then there is Heisenberg's uncertainty principle which says that the product of our uncertainty in where something is and our uncertainty in its momentum cannot fall to zero. Likewise, we can't know both when something happens and the energy of the happening.

This story will probably be classed as science fiction, but so far as I can tell from the equations of *our* physics, none of *the girl's* physics is fictional. And, with one or two obvious exceptions, none of the people and places mentioned in the story are fictional.

But let the reader beware!

John L. Dobson
Hollywood, March 2003

# THE MOON IS NEW
## Time Comes In With A Minus Sign

# John Lowry Dobson

Being a mirror wherein the physicist may see his physics from a different point of view, wherein the Vedantin may see his philosophy against the background of its physics, wherein the evolutionist may see genetics through a different pair of eyes, and wherein the reader may see his life against the final aspirations of his mind.


## INTRODUCTION


*Earth's crammed with heaven,*

*And every common bush, afire with God,*

*But only he who sees takes off his shoes. . .*

Elizabeth Barrett Browning


Similarities between the characters and happenings in this story and characters and happenings in the real world are *by no means* coincidental.

## ACKNOWLEDGEMENTS

I should like to acknowledge my extreme indebtedness to all those who have made this story possible, debts too numerous to count, debts too deep to describe, debts impossible to repay.  And, most of all, I should like to acknowledge my total indebtedness to that grand old man of science, Albert Einstein, in whose mind first bloomed the flower of modern science, and it is with the deepest humility that I beg of him, as he begged of Newton: Einstein, forgive!

## DEDICATION

And it is with a deep sense of gratitude that this story is dedicated to Swami Ashokananda whose instructions to me suggested it, to Richard Feynman who lacked the opportunity to read it, to Maria Berbeo who made it possible, to Sandra who cried over it, and to 'The Girl in the Front Row.'
But really, I suppose, it's dedicated to the Old Man in J.D.'s shrine.

CHAPTER 1


## THE FIRST DAYS: THE FOREGROUND STARS: THE MOON IS NEW


Mao had died the year before and, for the first time since the days of the Long March; I was in Peking when I received the crucial cablegram from Avila: "At last we've taught her English." The cable came from Berkeley with a date a few days before Christmas. But this was not news in the bleak Peking winter, where living and dying came first, and where memories of earthquakes still troubled the mind. Only to me would this be news, there on the frozen ground where I read it, where the wind was as cold as in the days of my boyhood, and where the persimmons were frozen and stacked in the streets. Only to me would this be news, for no one around me knew of this "girl," nor of why we had taught her our English.

The cable had come like a sudden spring to the frozen arctic tundra, and my immediate thought was to cut short my stay and hasten back to Berkeley. But I knew, if I did, that the news of this "girl" would leak out, and neither Avila nor I had wanted that.

We had kept the whole thing quiet from the start, and meant to keep it so. We didn't know from where she had come, and our only hope to know was to teach the girl English. We had tried before I came away from Berkeley. Apparently we failed. Now, where Avila and I apparently had failed, Avila, it seemed, had succeeded. That was the significance of the words "At last," and I was happy. But because of the strangeness of the circumstances under which we had found her, my mind was overwhelmed with curiosity about what the girl might say, and it was, therefore, with a sense, to me, of tremendous self restraint that I continued my scheduled stay in that beautiful city of my birth, without giving any sign, to my companions, of the delight or of the turmoil that raged within me.

Being a native guest of the Central Flowery Republic, as my father used to say, and fluent in its language from my youth, I had accepted an invitation to deliver a series of lectures at several universities in and about Peking. The lectures were to be on two separate subjects: Observational Cosmology and The Construction of Large (Ground-Based) Astronomical Telescopes. I had turned down Allan Sandage's suggestion, made the summer before, that I should go to Russia to help with their two hundred and thirty-six inch telescope. He had said that they were having trouble with it and had failed on three mirrors. Had I known Russian I might have gone anyway, even though my primary interest was no longer in telescope construction. My primary interest, now, was in cosmology and the concepts underlying our physics.

It was thus that, in the autumn of '78 (more than half a year later), with an unmassaged and burning curiosity, I again arrived in Berkeley and once again laid eyes on that marvelous girl. At first I was taken aback at how very well she spoke, as I had been gone for just over a year, and before I had left she had not yet spoken a word. Yet to hear her speak now, one might think she'd been raised in Berkeley. How is it possible, I wondered, that one, in so short a time, could learn to speak so well?

Looking back now I see it all, how little we knew and how vast were her talents, but in those early days we had no way to know. I had left Berkeley before she had spoken a word of English, and neither of us knew from where she had come. Who could possibly have guessed?

Avila had written to me, a month or so before my return, that they had toured Mount Palomar and Mount Wilson, and of how shocked she had been by the girl's response to the photographs of the galaxies, particularly those of our Local Group, as if they had been familiar objects to her eyes. Some of them she had addressed with endearing words, in her own tongue, much as one would address one's long lost dog, turned up in a strange, unpeopled wilderness. Then she had asked, with marked surprise, "What are the speckles all over the pictures?"

It's all so clear to us now, but no one knew then where she had been.

No one knew then of the peripheral stars, pulled from the disk of our Milky Way by the tides of other galaxies. No one knew then that a planet could roll with no stars in its sky but its sun. No one knew then that on such an earth the adaptation of a 'human' eye to darkness would reveal the interrelationships of the local galaxies, much as our own eyes reveal the constellations among our local stellar populations. And no one knew then, as no one knows now, how a life could survive in the cold, outer dark between the stars.

We had found her one day in the winter of '76, several feet deep in a snow drift, high in the Sierra. The spot, from a distance, had looked like a splash in the snow, and we would just have overlooked it had I not remembered reading, once, of a pilot who broke his leg forty feet deep in such a drift, high in these mountains, when his parachute failed to open. Quickly we approached the hole and peered in. Something under the snow looked red. I gasped and we began to dig.

In the days of my youth I had dug out my very best friend from a cave-in, where the Sunset Boulevard now runs in San Francisco. Jack Aratta and I lived on Thirty Seventh Avenue near Golden Gate Park, and they had taken out the buildings between Thirty Sixth and Thirty Seventh Avenues to put the Sunset Boulevard through there. And we kids had tunneled under the old basement floor of one of those old buildings. We had a little 'club house' in there with a kerosene lamp. And I just happened to be looking out the front window of my mother's house across the street and saw my youngest brother jumping up and down on that floor, and it caved in. I knew that Jacky was underneath, buried in the sand under that concrete slab. I panicked and I ran. My brother and I dug around in the sand till we found his hand sticking out from under this great slab of concrete which I had no idea we could move. But with all our strength, which was considerable, we lifted one edge and got Jacky out and he had coughed up blood and sand. If I hadn't, by accident, been looking out the window, we'd have dug him out dead.

I had panicked then, and I panicked now, and hastily we dug her out. She looked as though she had been thrown from a motorcycle into the snow, with her helmet still in place, but under the silver visor the face plate was red, and through it we could see her unanguished face, in the peace of sleep, or of death. We couldn't tell.

She was soft, light and easy to carry and there was no trace of rigor mortis in her limbs. As quickly as possible we carried her to the car, raced the engine, turned on the heater and drove for the cabin where we could build a roaring fire. We laid her out in front of the fire before attempting, in any way, to disrobe her, which turned out in the end to be a formidable task. The helmet was fastened to her outer garment with airtight fasteners, the likes of which we had never seen, and when finally we released one, there was a slight hissing sound as if she had been packaged under pressure. How stupid we were. How little we knew. How little we guessed. We had let some air out of her suit.

She was cold to the touch, but not icy, when we got her undressed, and we rolled her in front of the fire, taking turns at heating bath towels over the flames with which to handle her, lest we should cool her with our own cold hands. In the absence of a clearly discernable pulse it's hard to know what possessed us to roll for so long something which acted so much like a corpse. Perhaps it was the absence of any stiffness in her limbs. Perhaps it was the recollection of something we had read, long ago, that the temperature of the human body could be dropped to about $60^0$ F without harm by immersing it in crushed ice. I simply don't know, but we rolled. We pumped her lungs and rolled. When our hands became warm we massaged her limbs to increase the circulation of her blood, and before two hours had passed she moved and Avila whispered, "She's *alive!*"

Her first movements were very slow, like the residual motions in the body of a decapitated snake, but with them came unmistakable indications of a pulse, which soon were followed by her own efforts at breathing, great, slow, intermittent sighs, like the first, great blasts of steam in the railway engines of my youth, when the great, weighted wheels began to move.

As soon as she began to breathe we sat and watched. She was fair to look upon, well figured, small of stature, slender of limb, with a head a *bit* disproportionately large, like a child. As she lay there, slowly moving, with her eyes still closed, Avila remarked, "Her breasts are even smaller than Tina's." (Tina was a girl whom I had met at my brother's lake near Mount Shasta, and whom Avila and I had taken hiking on Mount Tamalpais later. There we would often hike, or sunbathe, without shirts, and, seeing Avila's well formed breasts, with roundish nipples, like the terminal knobs atop the Buddhist stupas, she had remarked about the smallness of her own, and the embarrassment that she had felt on that account.) "Yes," I said, remembering that that had been, with Tina, a sore point since puberty, "but look at the size of her nipples." In a way her breasts were more like those of a dog with pups; the milk bags were small but the nipples were obviously for business, with large, dark brown areas round about. The rest of her skin was lighter brown, like the skins of southern Asia, and when finally she opened her eyes she looked strangely and startlingly beautiful, as well as quite vibrantly alive, and the impression quickly rushed through my mind that if she were to stand before a large concourse of people no one would be able to turn his eyes away.

We sat there stunned, not knowing what to do. For the past two hours we had been totally occupied with the problem of keeping her alive. Suddenly now, that problem seemed gone, and, almost embarrassed, we sat there fully clothed, looking down on that beautiful girl, lying naked on the floor between us, and looking up with wondering eyes at one and then the other.

We spoke to her, though neither of us thought for a moment that she would understand. There was something about her so utterly different from ourselves that it seemed impossible to believe that she might understand our language. There was no response. She looked at one and then the other with those large, unblinking eyes as if no word had just been spoken. We touched the various parts of her body to see if she felt pain. At no point did she wince or show the slightest sign of shyness. If we rolled her she would not resist, but she would not roll herself, nor would she sit.

My first thought was to feed her something warm. But what to feed her? And I was afraid. Ernest Thompson Seton advised that when you capture a wild animal don't offer it food until it's starving, lest it turn away from food and starve itself to death. Who was to know? We felt it better to withhold food and drink, and watch her closely for any sign of distress.

She lay by the fire all that day, covered with a blanket, while we took turns at fetching wood. We ate our evening meal there beside her. She watched but never reached. Her motions were unbelievably reserved, like one who's just been missed by death, or one about to die.

That night became a worry to us both. Why had we not gone for help?

What if in that long, dark night her life began to fail? We lay beside her on the floor in fitful sleep, listening for any sound, and all too frequently we tested the temperature of her hands, feet and forehead.

It seemed the dawn would never come. The fuel was getting very low. We talked of bundling all together to keep her warm. We didn't know what she would think, what she would do. Her breathing seemed more normal now, which eased our minds a bit, but before the dawn the room was cold and we both crept in beside her, one on either side, with the sleeping bags spread out as far as they would reach.

In the morning she seemed stronger, as if she were coming back from death. We ate our morning meal, putting everything within her reach. She never made a move, but when we were through, she smiled. Ever so faintly she smiled.

By evening she sat up and tried to eat. Tenderly, at first, she tried, in tiny bites. That night we were happy and slept together on the floor with the bags spread over the three of us, like cord-wood in the rain. The joy of sleeping by a stranger, that joy I had, that night, in the extreme.

After she started eating, her strength picked up quite quickly, her limbs filled out a little, and we brought her down to Berkeley where it's not so cold.

By late February it was sometimes warm and sunny and we took her to the hills where we look down on all the towns laid out around the Bay. We talked to her. We pointed to the sky and named it. We pointed to the Sun and named it. We pointed to the Bay and named it. We pointed to the towns and named them. We pointed to the trees. We pointed to the birds. We pointed to each other. We pointed to a deer. There was never a response. Once she smiled, when we pointed to the Moon, and once again at night, she smiled when we pointed to the stars, but never a word, never a sound.

How little we knew! How little we guessed! It never crossed our minds why she smiled at the Moon. The only reason I even remembered it was that once I had a dream, one of those beautiful, heavenly dreams, in which one of my favorite people, Jo Stanbury at the Vedanta Society (everyone called her Jo), said of me, "He can smile at the Moon but still see through it." That dream left such a vivid impression on my mind that it's vivid even still and was at once recalled to mind by the smile of that girl.

Those were sweet days. Whatever relationships we had with that girl were wordless. But gradually we came to doubt that she even had the ability to speak. Then came the diarrhea, that long siege of diarrhea, which reduced her almost to the state in which we had found her. By the last day only her eyes and her bowels would move, and we thought surely she would leave us. Then, unexpectedly, next day, she sat up and took her food. And then, within a week, I left for China.

It was, therefore, with great surprise that I found, after an absence of just over a year, that she had, by now, become more fluent than myself in my own language. And although I had been informed, by mail, of her accomplishments, I was not at all prepared for what I found. Not only had she become quite competent in speaking but also in reading, and at such great speeds that by the time of my return she had all but exhausted the reading resources which she had found interesting in our small library.

Our small library, Avila's library really, for all its smallness, was heavily weighted to mathematics, physics and astronomy, for I had brought with me all my books when I moved in. All these things she loved, and all these things she read. But I must point out that, from the frequent conversations which ensued between us in the following weeks, the notion gradually insinuated itself into my mind that the contents of the math and physics books were known to her before. Never for a moment did she imply that it was so, but it became ever more apparent that it was.

It was in astronomy alone, which intrigued her most, that she seemed to be reading what, to her, was new. One day I saw her looking at the pictures of galaxies in George Abell's book, *Realm of the Universe,* and, remembering what Avila had written about the trip to Palomar, I asked the girl about the "speckles" on the pictures. Right away she smiled and said, "We don't see these foreground stars." Taken aback, I regrouped my thoughts and put forward my next question with hesitation. "You mean that you're familiar with the galaxies but without the foreground stars?" "Yes, exactly," she said, "without the foreground stars."

Then, for the first time, something blossomed in my mind. I remembered how, before she spoke a word, she had smiled once, when we showed her the Moon, and once, when we showed her the stars. "Is the Moon also new to you?" I asked. "Yes," she said, and smiled, "We have no moon."

"Where do you come from?" I asked.
"I do not know. I look through these books to find out, but I do not know."

For several days I was busy, out of town, and couldn't see her, but often I remembered what she said. I remembered almost every word, for by now I knew too much to take her lightly. When I got back, I asked her, "Do you see those galaxies with your bare eyes, or with some sort of machinery?"

"With telescopes," she said, apparently amused at my clumsy choice of words, "but many can be seen with our bare eyes."

"With telescopes?" I asked.

"Yes," she said, "but not like yours." And she explained to me how their big telescopes were built.

What she sketched and described was a lovely device, a floating device, like the twelve inch telescope which Norman James had brought to the Riverside Telescope Makers Conference in Southern California a few years ago, and which beat out the Sidewalk Astronomers, twenty-four incher in the judging.

Norman James's 12 incher

But, unlike his telescope, the telescopes which the girl described had the rest of the telescope separated from the tracking device, so that the telescopes could be scaled up to almost any size. That's what charmed me.

The objective mirror lies face up, and motionless, in the bottom of a pit, over which stands a tower, open on what here, in California, would be the southern face. In the top of the tower hangs a tilted flat (mirror), face down.

South of the tower, on the ground, and down the polar axis from the upper flat, floats the tracking flat mirror. The tracking flat tracks a position in the sky midway between the target and the celestial pole and sends the reflection up the polar axis to the upper flat and down to the objective. The returning (and converging) beam focuses, after reflection by the upper flat, on the center of the tracking flat where the "sensing devices," as she called them, were placed.

Don't you see the fun of it? The machinery is separated from the objective. Only the tracking flat is out under the sky and in need of protection from the weather. Smart. All the moving parts are out there.

The tracking flat, which forms the upper surface of a hemispherical boat, floats, like Norman James's sphere, in a basin. The drive shaft comes up the polar axis to some sort of magnetic attachment to the tracking boat. (Norman James used an O-ring.) The drive shaft makes only one revolution per day and tracks the object across the sky by tipping the boat.

She sketched it on the sidewalk, and I have tried to reproduce her sketch from memory. It was too impressive to be forgotten.

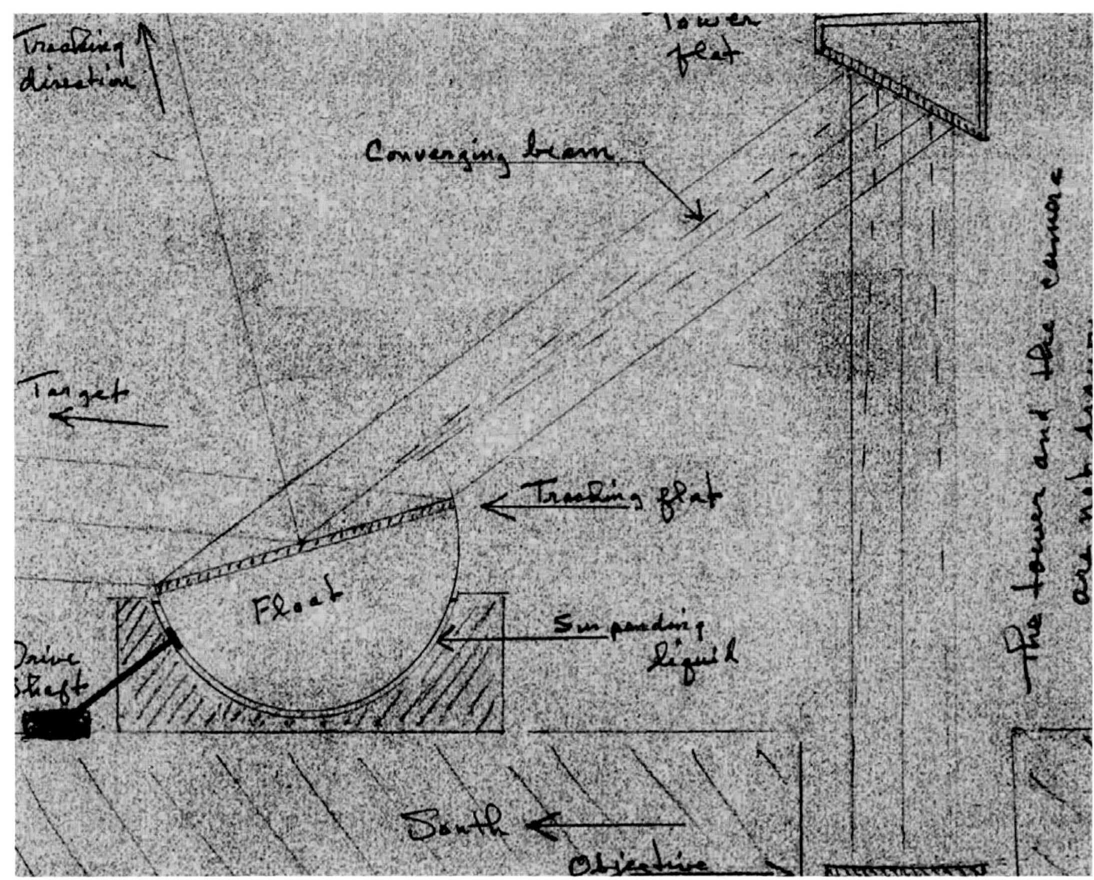


My sketch

The advantage of this construction, she pointed out, is that it can be scaled up to virtually any size without running into the cumbersome problem which she had noticed at Palomar, namely, that as the size of the instrument is increased, the tracking device becomes so unwieldy that one is tempted to shorten the focal length, unreasonably, and thus curtail the effectiveness of the instrument.

There were two other advantages that she mentioned, which I still remember, 1) its freedom from the turbulence problem arising from the enclosed body of air in the dome (as at Mount Palomar and Mount Wilson) and 2) its freedom (or comparative freedom) from the deterioration of the image clarity due to the optical obstruction of the secondary mirror. She said that the central obstruction is held to a minimum by keeping the "sensing devices" (camera, spectroscope, etc.) small, and that the air turbulence due to the cooling of the legs of the tower "by infrared radiation" is controlled by "maintaining its exposed surfaces at the ambient air temperatures." Where, I wondered, has she picked up all these words?

She said that they have very much better "seeing conditions" at their observatories than we do here and can distinguish some individual stars in the globulars of the galaxy which we here call M87. She further said that the sky there is very dark, and probably, as she suggested, since there are no stars or moons in the night sky, the adaptation of their eyes to the dark is much more acute than ours; so that no one would dream of lighting up a town as we do, or of shining a light up into the sky at night. She mentioned also that there is almost no smog there except after volcanic eruptions, which, she said, tended to be "more like your Krakatao," or during the fire season. They burn their forests selectively, "just before the rains," so the rains clear away the smog. And she said the internal combustion engine is almost unused. Most of their energy is from windmills wired to fly-wheels. Only just before the rains does the smog get bad, but there is nothing much to light it up. She said that even in their "Festival of Lights" they wouldn't dream of lighting up a town the way we light up Berkeley, though she did admit that it's pretty from the hills at night.

By now it was May and we took her for walks in the hills in the daytime. There, if she were seen, she'd be seen at a distance, and, because of her small stature, could easily be mistaken for a child. The startling appearance of her eyes was greatly subdued by Avila's green glasses, which she had worn to Palomar, and, although her voice was unusually musical and pleasing to the ear, it would not be likely to attract attention unless the words were overheard.

There was, about her, both in her physique as well as in her face, a look of extreme refinement, and the expressions of her face, as well as the nuances of her voice, showed such a wealth of variety as to challenge one's power of interpretation. During the first days this had given rise to some confusion in our minds since her use of these expressions did not, at first, correspond exactly to our own. Later this confusion was to melt away, whether by our becoming more accustomed to the way in which she used these expressions, or whether by an adjustment, on her part, in the use of them, to correspond more closely to our own, I cannot say, but her smile, which to us, expressed such a friendly openness and such a touching sweetness as to be at once understandable from the start, was used very much as we use ours.

Soon it would be June and we could get away together, all three of us, maybe to the mountains, or some lake, where we could talk. Maybe we could take her up to Lick Observatory, to use the telescopes, or find out where the Sidewalk Astronomers would be. At night in a crowd she could easily go unnoticed.

When first we picked her up, out of the snow bank, we were concerned about the safety of her life. When first we brought her down to Berkeley, we were concerned about the safety of her mind. What if 'Missing Persons' couldn't find where she was from? We thought it better first to teach her English and find out. Now that she knew English it was becoming more and more apparent that she was of no concern to 'Missing Persons' and that, at any rate, they'd never get her home.

# CHAPTER 2

## THE LONG CONVERSATIONS: E = M

Summer saw the beginning of a series of long dialogues which now I wish we had on tape. They started on the day we left for my brother's lake near Mount Shasta. Avila was driving and I sat in back, and once we were out of the city and up highway I-80 in the clear, but long before we got to I-5, the girl gave up peering out of the windows, and, looking over her left shoulder, turned to me with a smile and said, "You have a famous equation, one of Einstein's famous equations, $E = mc^2$. What does it mean?"

Since, in all these weeks, she had never started such a discussion before, I didn't know what to make of it. I didn't know what to expect. It was her habit not to speak until spoken to. I don't know why, but she always did it that way. I didn't know what I'd be in for; so, in order to avoid leading the conversation, I asked her, "Why do you ask?"

"Well," she said, playfully, "in your college astronomy texts I find an odd thing. In the Jastrow and Thompson astronomy textbook, on page one hundred and forty, they quote Einstein's famous equation, $E = mc^2$, and then turn right around and say that matter can be *converted* to energy, that the *sum* of mass and energy is indestructible. That would be $E + m = K$, but there is no plus sign in Einstein's equation. Am I to think that Jastrow and Thompson, and I could name others, are misinterpreting Einstein's equation, or am I to think that Einstein was a fool?"

"Surely the former," I said. "Those who invent good physics are few, while those who teach physics to others are many, and do not *always* understand."

"I knew you would say that," she said, "but I had to ask. Einstein's equation says that mass and energy *are the same thing*, and, I suppose, the $c^2$ must be the number of ergs that make a gram."

"Yes," I said, and I turned to Avila to explain. I told her that in the centimeter-gram-second system of units, which was in use in Einstein's day, and in my day, the unit for energy was very small and the unit for mass was very big. And when Einstein found that they both measure the same thing he had to know how many small ones were equal to one big one. He had to know how many ergs are equal to a gram. Energy was measured in ergs, and mass was measured in grams. And the *erg* is the kinetic energy of a beetle, weighing only two grams, walking one centimeter per second and running into your shoe, while the *gram* is the energy of an atomic bomb. The c squared is just how many ergs are equal to one gram. And it's a huge number.

Then I turned to the questioner.

"One thing I learned at the University is how to read equations, and, as you say, there's no plus sign in Einstein's equation. Einstein, himself, referred to that equation as, 'The equation in which energy is set *equal* to mass.' However, it's usually taught as Jastrow and Thompson have it, that the *sum* of mass and energy is indestructible."

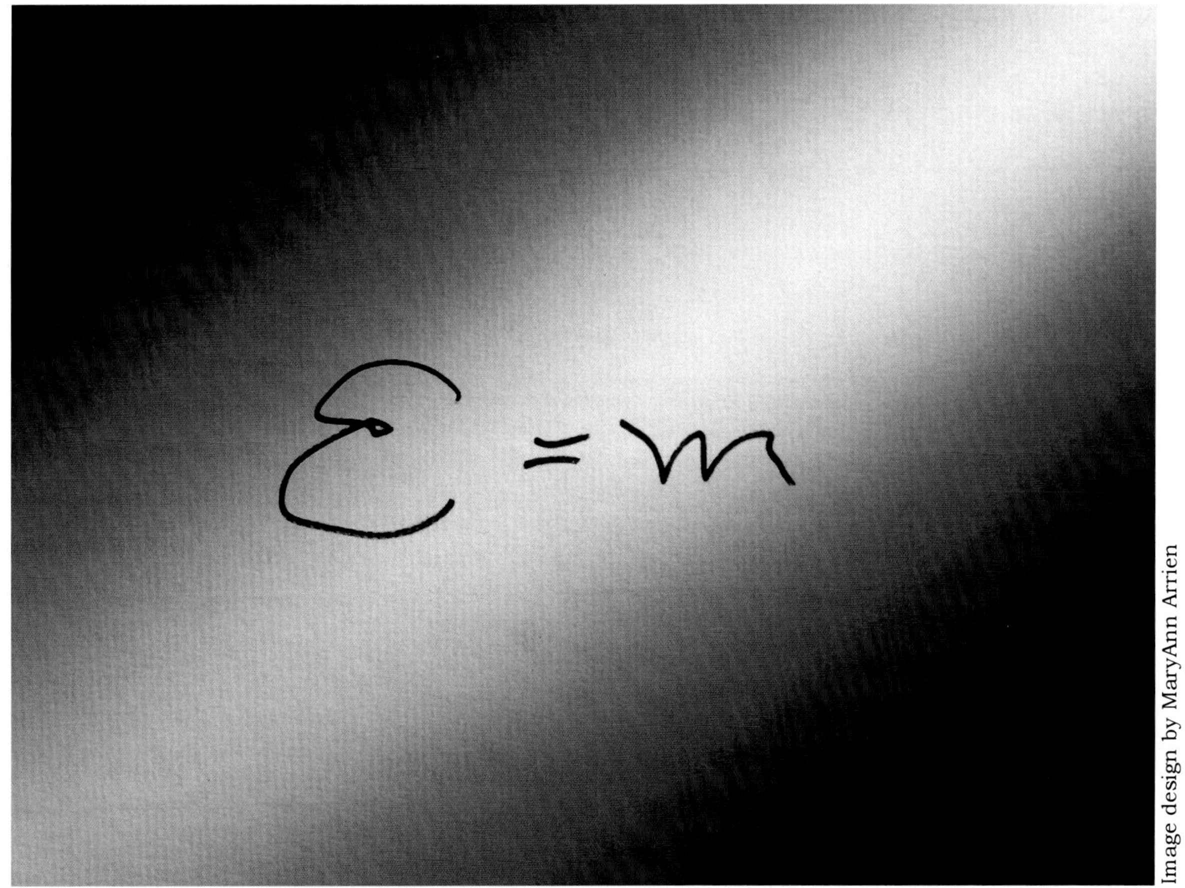



Einstein's physics, 1905

"Then, why didn't Einstein clean up the academic community?" she asked. "He lived here till nineteen fifty five."

"Probably," I suggested, "because no one in academia would have *dared* to teach relativity *in front of him.* He may never have seen how it was done. And in their papers the physicists would have used *his* equation, not E + m = K."

"Surely you're right," she said, "But how did *you* get into astronomy and all this telescope making?"

"Well, really the telescope making came first," I said. "That was in the monastery. But before I joined the monastery I had a degree in chemistry, and any chemist can see what Prout saw in eighteen fifteen, that the rest of the chemical elements must be made out of hydrogen. And by the nineteen forties I think we *all* knew that the Universe is still mostly hydrogen, and anyone could see that it's running on gravitational collapse. I just wanted to watch it. I wanted to see what the Universe *looks* like.

"But, you know, Swami Vivekananda wanted monks to go from village to village with Magic Lanterns, an obsolete ancestor of our slide projectors, to educate the villagers in science. That's really what I was doing with my telescopes in Sacramento, and that's what the Sidewalk Astronomers do now. We run through the national parks and Indian Reservations with slide shows and telescopes.

"But making our telescopes started in the monastery in San Francisco when Bill Wuebold told me that you could grind your own glass. And we started on Bob Reed's twelve inch porthole glass. Then, when I got shifted to Sacramento, and didn't have any big glass, I made five and a half inch telescope mirrors out of the bottoms of gallon jugs. The *old* jugs had flat bottoms. I had to grind them with granite sand which we bought by the ton for the garden. And I had to make my own screens, screen my own sand, make my own rouge (for polishing compound), and silver my own mirrors. And nowadays when people have trouble in my telescope making classes, none of my tears come down."

That made her laugh, and I explained that I had to cut out the bottoms of the gallon jugs, and that I ground them against a six inch cast iron stove top under water. She thought it was very funny that I ground them in a pail of water so it wouldn't make so much noise; so I explained that making telescopes was not on our curriculum.

"How did you know how to make rouge?" she asked, "And out of what did you make it?"

So I told her that chemists know how to do such things. And I told her that we already had oxalic acid for bleaching wood, and ferrous sulfate for fertilizer. I told her that I precipitated ferrous oxalate, a beautiful yellow powder, and that I washed it and dried it and put it under a flood light.

"Under the flood light," I said, "little fires ran through the pretty yellow powder like the fires that run through the soot at the back of the fire place, and I thought I had wrecked it. But what looked like soot was rouge."

"You're pretty smart," she said, "What about silvering? Where did you get the silver?"

I told her I didn't remember that part but that I made silver ammonia complex ion from silver nitrate and ammonia, and that when I threw in the reducing agent the silver precipitated all over the mirror and everything else in the container. But I told her that silver mirrors get tarnished easily and that nowadays we get our mirrors aluminized and overcoated in a vacuum chamber, and that the aluminum coatings last for some twenty years.

"But it wasn't these five and a half inchers that got you into trouble." she said.

"No," I said, "It wasn't. Buck Tergis who came up from San Francisco to work in the garden saw Saturn one night through one of my five and a half inchers and said, 'J.D., you've got to make something bigger even if I have to dig you an underground workshop.' So I asked him to go to the marine salvage shop at the foot of Filbert Street in San Francisco and send me a list of the porthole glass sizes. That's what he did. And since I knew that I could get twelve inch inside diameter cardboard tubes twenty-two inches long from the rubber companies, from the cores of hose reels, I asked him to get me some twelve inch porthole glasses."

The problem then was to send them to Sacramento without having them discovered by the people who carried them; so I told her that Buck opened the hundred pound bag of ammonium sulfate, which we needed for fertilizer, and poured it into four cardboard boxes with twelve inch porthole glasses in the bottoms. And she laughed when I told her that the people who carried them never understood how that hundred pounds got so heavy.

When she heard that the cardboard tubes were only twenty-two inches long she said, "That's too short." So I told her that we glued them together in threes and fours for sixty-six and eighty-eight inch focal length telescopes, and that we called them 'three-barrel twelves' and 'four-barrel twelves.'

"So *that's* why you made those two kinds of telescopes, sixty-six and eighty-eight inch focal lengths," Avila said "to match the lengths of your tubes."

"Yes," I said.

Since we didn't have much money in the monastery, the girl wondered where we got the wood for our rockers; so I told her that the mounts for our telescopes were made from the discarded window cut-outs from school house doors and that they were two by three feet across and nearly two inches thick and that we cut them in two for the side boards, the front boards and the bottom boards of the rockers. I told her that the shingles for the spider mounts to hold the secondary mirrors at the front were blown off the buildings in the wind, already blackened by the sun, and that we didn't have to pay for any of this.

"But you didn't build the twenty-four incher in Sacramento, did you?" she asked.

"No," I said, "That was later at Fourth and Anza in San Francisco where Brian Rhodes lived. He had come to me for help in figuring his eight incher, and then he made a sixteen incher which we took to the second Riverside Telescope Makers Conference in nineteen sixty-nine."

(The Riverside Telescope Makers Conference was started in 1968. And they heard about the Sidewalk Astronomers in 1968, which was started that same year. And they invited the Sidewalk Astronomers to the Riverside meeting in 1969, where they tied for first place in optics, and were runner-up in mechanics even though their mounts did not track objects across the sky for photography.)

"After that, Brian and I made Delphinium (the twenty-four incher). We made it freeway portable, in three months, for three hundred dollars. But we didn't have a vehicle big enough to haul a twelve foot tube, so we wanted a four foot piece and an eight foot piece of thirty inch tube. And when I went down to Burke Concrete Accessories to order it, I found a four foot piece and an eight foot piece of thirty inch tube standing on the loading dock. And I said to Brian, 'Did you come down earlier and order this?' 'No,' he said, 'Didn't you?'

"It was dunnage from someone else's order that morning, standing on the loading dock, and I said to Brian, 'Something in this Universe wants that twenty-four.'

Brian Rhodes, Joannie Jackson and me with the twenty-four incher at Riverside
Our fail-safe sun telescope is in the background

"But," I said, "You asked how I got into astronomy. That came later in San Francisco when I was teaching classes in telescope making. One of the kids with a twelve incher, and his school teacher, asked me to teach astronomy. That's how *that* started."

"But I thought you never studied astronomy," she said.

"That's right," I said, "I never studied it *at the University*, but running around with telescopes I was exposed to a great deal of astronomical information, and if you're interested in the information, you pick it up. So when that kid and his teacher asked me, I thought, 'Why not?' so I did it. And it was in those classes, later on, that I ran into the significance of E equals m."

"And really," she said, "I suppose that E equals m got into European physics through Swamiji (Swami Vivekananda). Do you know *how* it got in?"

"Yes," I said, "Swamiji met Nikola Tesla at Sarah Bernhardt's party in

New York in eighteen ninety-six and asked him if he could show that what we call matter (mass) could be reduced to potential energy. Well, apparently he didn't get it done, but when I was in Tasmania, I mentioned all that to the audience, and I pointed out that, some ten years after that, Einstein had gotten it done. When I finished my talk, a young man, descended from Tesla's brother, told me that they have it in the family archives that Mileva Maric, Einstein's first wife, who wrote the papers for her husband in nineteen hundred and five, was a close friend of Tesla's. So she would have known the problem. That was when I first saw the connection between Swami Vivekananda and Einstein which had puzzled me for years.

"Mileva, of course, was a physicist and when she wrote the papers for her husband, she would have noticed the connection between their relativity paper and his quantum mechanics paper. In their *relativity* paper he had pointed out that if something is seen to be receding, its *mass* will be seen to be lowered. And in his *quantum mechanics* paper he had pointed out that if something is seen to be receding, its *energy* will be seen to be lowered. (It's only if something is seen to be *approaching* that its mass and its energy will be seen to be increased.) There is no way that she wouldn't have noticed that the mass and the energy go up or down in the same way. Mileva has not been credited for her contributions to relativity theory. E equals m is surely hers. The c squared is probably his. When Einstein found out that ergs and grams measure the same thing, he had to know how many ergs make a gram."

"So Swamiji himself tried to fix the physics," she said.

"Yes." I said. "He did, and with Mileva's help he succeeded. And I might be the only one on the planet who knows *how* it happened. But *very* few physicists have noticed. As you see they don't take his E = m the way he wrote it. "

Avila knew, of course, that I had spent 23 years in the monastery of the Ramakrishna Order under Swami Ashokananda, and both she and the girl had access to my Vedanta books which I knew that curious girl must have read. I had, among other things, a copy of *The Gospel of Sri Ramakrishna* by "M" (one of his disciples, and the Head Master of a school) and some of Swami Vivekananda's writings. And both of them knew, of course, that the Ramakrishna Order had been started by Swami Vivekananda (whom we call

Swamiji) after Sri Ramakrishna had passed away in 1886.

Near the famous Nut Tree restaurant we got off of the highway for gas and to stretch our legs. It's a little way before the turnoff to I-505, which takes us north to I-5.

When we got back in the car I felt that Avila needed an explanation, so, addressing her, I said, "Swamiji had the job of presenting Sri Ramakrishna's Advaita - non dualism - against the dualistic science of the West, and against the dualistic philosophies of India. And Swamiji spent his best teaching years in the West, where we think against a background of science and not against a background of philosophy as they do in India. And the European science of Swamiji's day was dualistic. It had the Universe made of *two* things, both matter *and* energy. Swamiji wanted just *one* thing, *just* energy. And I think that's why he asked Tesla if he could show that what we call matter could be reduced to potential energy. And Tesla even used the Sanskrit terms *Akasha* and *Prana* for mass and energy.

"But," I said, turning to the girl, "Swamiji had another problem. He had to figure out against what philosophical background could we best understand Sri Ramakrishna's life and teachings, whether it was dualism, qualified non-dualism, or non-dualism. Advaita is non-dualistic Vedanta. Most of Christianity is dualistic; God and the soul are thought to be separate. Most of Vedanta is non-dualistic; there is no such separation. But there is an in-between position in Vedanta where the separation is thought to be less.

"Once, when Swamiji was giving a talk in India, he mentioned that these are just different points of view, and that the religious aspirant goes from dualism through qualified-non-dualism to non-dualism. They are just different points of view, but of the same reality. Then someone called out 'If that's the case, why didn't the old Rishis say so?' Then Swamiji said 'I was born for that, and it was *left* for me to do.'

"And it was Swamiji who had the eyes to see that everything Sri Ramakrishna said or did was based on Advaita, non-dualism. So, in spite of the fact that we read in M's Gospel that Sri Ramakrishna said, over and over, that so long as one has body consciousness one should not say 'I am He,'

Swamiji taught Advaita all over India and the West."

"Surely you're right," said the Little One, "How come I never read all that?"

"Because it's not written," I said, "That was from Swami Ashokananda's morning class in the monastery. He left us for a whole week with the problem of reconciling what Swamiji did, with what we had read in M's Gospel. We got nowhere with it. Then he told us that M was a householder (a non-monastic) and that in his Gospel we read what Sri Ramakrishna had taught to the householders. And Swami said that that's not what he taught to the monks, who said that M's Gospel was like a curry without salt. Then he said what I just said, that Swamiji saw that everything the Old Man had said was based on Advaita. And, finally, Swami told us that when someone of Swamiji's stature teaches something, he makes a change in the Cosmic Mind, and, after that, *that's* the way to do it. So *now* it's *Advaita*."

Swami Ashokananda

Then the girl continued.

"When I'm reading what you call physics," she said, "I read one thing, and when I'm reading what you call astronomy, I read something else. It's true in several instances. For example, your astronomers regard gravity as all important while your physicists think it's nowhere."

Suddenly the injustice of my treatment of that girl overwhelmed me. Till that moment it had never crossed my mind that she had studied all our knowledge from those stupid books in our library, and that in one short year she had mastered it and judged it.

"I'll get you better things to read," I blurted out. "You should read Einstein in the original." I paused. "But he wrote in German."

"I can learn German," she said, with casual assurance. "After learning English, German would be easy. It's the first language that's hard, you know, because you don't know how they think. You all think just alike. You have no idea how just alike you are."

She went back to peering out of the windows, but by now my mind was reaching. What does she know that we don't know? When I had asked about optical machinery she had given me a new design for telescopes which probably would not have arisen in a land where the turbulence was bad nor in a land where the telescopes were small, and she had told me that they could discern some individual stars in the globular clusters of the galaxy we call M87. I needed time to think.

Avila knew, of course, that I had been born and raised in Peking, China, and that my brothers and I had been brought to San Francisco by our parents in 1927. At that time I was twelve years old. My mother was a musician, also born in Peking, and my father was a Canadian zoologist. He had gone 'out to China' to teach zoology at the Peking University which had been founded by my mother's father, Dr. H. H. Lowry, who had been Advisor to the Parliament of Religions at Chicago during the Columbian Exposition in 1893.

It was at that Parliament that Swami Vivekananda had suddenly attained to world renown, much against his wishes, by captivating those audiences with his ringing words. Swamiji was a prominent disciple of Sri

Ramakrishna who had passed away in 1886.

Probably it was due to the influence of Sri Ramakrishna and Swamiji upon my mother's father that it had been taken for granted in our home, throughout my youth, that all religions are true, that is, that religious experience is attainable and that there are many paths by which to reach it. That is not to say that all *philosophies* are true. That's a different matter. Philosophies are true or false according to whether or not their teachings correspond to fact. But religions are *paths* by which to reach a goal, and if that goal is true, and if, by that path it's attainable, then we may say that that path, that religion, is true. Sri Ramakrishna had known from his own experiences that all religions were indeed true - that the goal is real and that the goal can be reached through many different paths.

But by my late teens, and through my early twenties, I had taken for granted that the reverse was true. I had taken for granted that the case was closed against religion. I regarded it as all but proved that God was a myth, and that 'saints' and 'holy men and women' though often very fine people, were certainly deluded.


Far from the Bay Area the weather was warm, so when we pulled off of the road for lunch, we sat under a bridge by the Sacramento River, and there we talked of other things. But she had raised a very interesting question, to which my mind repeatedly returned. What if her understanding of physics were totally dissimilar to our own? Our own understanding of physics has left many of the basic questions unanswered. Why does matter appear as discrete electrical particles showing gravity and inertia? Why are they electrically charged, and why do they resist changes in their states of motion? What determines the charge on the proton? What determines its mass?

Often I had wondered what it would be like to meet someone whose education had been completely different from my own. On what common ground could we converse? Judging from the question which she had chosen for openers, she had already understood the background of my education, and had thus made it possible, at the outset, for us to talk on common ground. This, I thought, promises to be interesting.


Later that day, on I-5, but long before we reached my brother's lake, or even saw the snows of Mount Shasta, it became apparent to me that she would not, on her own initiative, return to the earlier discussion. It was up to me. And I wanted to know how they saw things on *her* world.

"Do you have famous equations like $E = mc^2$?" I asked.

"Our ways are different," she said. "Famous doesn't apply, but equations we have."

"How is it," I asked, "that you have been able, in one short year, to learn English and then master all the physics and math to which you have been exposed, even to the point of picking out a mistake in the usual interpretation of Einstein's relativity theory? And how is it that you are so especially fascinated by astronomy?"

Suddenly I became aware of how totally rude it is to ask someone a second question without ever waiting for an answer to the first. I knew she would notice.

"I'll answer your second question first," she said. "In studying English, physics and math I had first to understand how you thought, how you saw things. I had to understand what was taken for granted in your thought and in your language. And I had to understand which of the things you had taken for granted are true and which are false. Once you understand what truths a man has taken for granted, and what mistakes, it's easy to size him up. You already know how he will talk. You already know what he'll say.

"You probably thought that for the first few months I didn't learn any English. That isn't true. I knew what you meant when you pointed to each other and mentioned your names. I knew that you were John and she was Avila. I knew what you meant when you pointed to the deer, to the sky and to the sea. I knew what you meant. I simply wouldn't answer till I sized you up.

"It was clever of you, though, to notice that I smiled only twice, when you pointed to the stars and to the Moon. That was very clever. I was pleased. 'Is the Moon also new to you?' you asked. You are really quite observant."

By now I felt like a school boy whose behavior has come under the scrutiny of his teacher. 'But what a sweet teacher,' I thought. Even when she makes fun of us, it's fun for us too. She was as loving as a mother and as innocent as a child.

She continued, "Avila also noticed, with surprise, that I recognized the galaxies but that I failed to understand the speckles on the pictures." She chuckled, then continued. "What you call physics and math is about the Universe in general. Math is just the logic of the mind by which we understand it, and physics is the way it behaves. But *your* physics is formulated from a peculiar standpoint, of course. Some things are taken for granted which aren't true. But much of what you call *astronomy* is about specific things, individual stars, the Solar System, galaxies and so forth. To understand your math and your physics I had to understand the point of view of your mind. To understand your astronomy I had to understand your point of view in space, and that's new. You seem to be living in the disk of a galaxy, and to me that's *new*. Nearby stars are new. The *Moon* is new."

"Why do you say 'what you call astronomy?'?" I asked.

"That's because we don't divide our branches of knowledge up in the same way," she replied. "You have chemistry, physics, biology, zoology, astronomy, philosophy, psychology and so forth. We divide in another way, how matter behaves, how life behaves, how mind behaves, how consciousness behaves if, indeed, one can speak of that, and then, what matter is doing *here,* what mind is doing *here,* and so forth. Chemistry, physics, math and some parts of philosophy are about the native behavior of things, but botany, zoology, and some parts of astronomy are about what is happening *here.* In reality, knowledge is not divided up like that at all. You used to call it all philosophy. Then chemistry and physics arose as parts of 'natural philosophy', and so on. One of your big mistakes is that you split your knowledge up like that."

Lordy, Lordy, I thought. This girl has analyzed everything we do. She looks and acts like an eleven year old kid. How could she know all that?

"To me," she went on, "the substance of your math and physics were not new, and only your point of view was a little new. But much of the substance of your astronomy is new. The sky doesn't look like this back home. And your botany and your zoology are new. Just as the flora and the fauna vary from time to time, and from place to place, just so astronomy varies from place to place, except, of course, that part of astronomy which deals with the fundamental nature and behavior of matter."

"Biology, then," I said, "particularly molecular biology, you would classify differently from botany or zoology."

"Exactly," she replied, "you've got it. Biology is the study of life processes in general, whereas botany is about how the eucalyptus does it in Australia and how the Pampas-grass does it in South America. Zoology is about how the kangaroo does it in Australia and how the horse, the camel and the saber-toothed cats did it in what you now call 'The States'."

"I understand," I said, "more logical. Do you know yet, where you came from?"

"No," she said, "but it must have been a star quite close to, but outside of the flattened disc of a galaxy - surely this one. We could see a close galaxy all across the sky, on one side of us, with a blue-shift from the stars, centered on a position rather far out on the disc, our side of center, as if we were headed for the outer regions of that disc. Are there disc stars, outside the disc?"

"I don't know for sure." I said. "When we get back we could ask. There are the halo stars, of course, pulled from the globular clusters, but they're older."

"No," she said, "It couldn't be an early star from a globular cluster or it wouldn't have planets. Only a spinning star could have planets. The angular momentum, of the original cloud that formed the Milky Way would have been transferred to the halo by what you call the great magnetic eggbeaters of the early stars. That's why the halo, what you call the hovering layer, was flattened into the disc. Only the disc stars have enough angular momentum to form planetary systems. Are there disc stars outside the disc?"

"It's not impossible that there should be disc stars near the disc of a galaxy if the spiral arms were formed, as we think they were, by tidal interaction with a passing galaxy. You remember the computer-generated pictures in Jastrow and Thompson? Probably you remember the page."

"I do," she said.

"Well, how did you get that way?" I asked.

"That was your *first* question." she said, and proceeded with the answer. "It's due to breeding - not in the way that you mean it when you say 'she's a lady of breeding.' I mean, we have it in our gene pool."

"You quote our ideas like an expert." I interrupted.

"They're not *your* ideas," she laughed, "but your *expressions*, I quote."

"What do you mean," Avila asked, "you have it in the gene pool?"

"I mean," she said, "that we select for child-bearing, breeding stock if you like, only those individuals who show a marked superiority in certain capacities and a marked absence of others. In order to be selected as a parent one must show outstanding manual dexterity and an outstanding capability at handling language and ideas. Handling ideas involves the ability to collect and store large amounts of information, the ability to sort and interrelate it, what you call intelligence, and the ability of recall, what you call memory. But, to be selected to make a contribution to the gene pool, one must also show a marked absence of responses of anger, hatred and what you call the sex urge.

"I knew you wouldn't understand." she continued, "No, we don't marry, nor anything like that. Men and women often become very fond of each other, and it is a very beautiful thing, but there does not arise between them what you call sex. Children are not born that way. Sex is thought of as an animal behavior, and, although sex between people is referred to in our ancient books as a past tradition, I have seen it here for the first time. Now, for the first time, I see what they rooted out. Forgive me!

"I knew you'd have trouble with this," she said, "but you must have noticed my behavior. You must have noticed that I never asked."

"We noticed," Avila said, quietly.

"You see," she said, "you are all descended from a long line of ancestors not one of whom failed to 'do it', as you say, but I am descended from a much shorter line of ancestors not one of whom *cared* to 'do it'. That's one difference between us."

"How, then, are children born?" asked Avila.

"By what you call 'artificial insemination'." she replied. "Although by now it seems quite 'natural' to us. The women, of course, must still bear children in the usual way. That has not changed, I still have breasts. Our women look very much like yours, except that our unused breasts are smaller, and some of us have more than two breasts. But the behavior of our men is very different. They no longer use the penis in the old way. We can always get offspring from a male without that because our men can discharge semen at will, much as your dogs can pee against a pole.

"You must have noticed that my breasts are small. They are not playthings for companions in sex. They are no longer selected for size and plumpness, and small breasts, to us, are not a cause of embarrassment, as they are to some of you. But the mammary glands are alive and well. We menstruate and nurse our offspring like you. Only the selection of our parentage is different. We tend to impose on our favorite people the task of parentage. It's hard on the women but not so hard on the men. The men can give semen on request and father as many children as we choose. *We* choose. *They* don't. The *women* choose."

"The trouble with *your* species is that you are too recently descended from the biggest bullies in the clans, what you call the alpha males. Long ago, in the jungle and at the beach, *they* fathered most of the children. And it shows in your *men,* more than in your women. But *now* your *women* also get to choose who fathers their children, as do *ours.*

It was time to quit talking, because Avila needed relief at driving. But this was the first of the series of long conversations, which I shall try my best to recount.

Avila, much later

## CHAPTER 3

## THE LAKE AND THE CANOE

The next long talk took place between the three of us in a canoe, drifting aimlessly in the warm sun on a small, cliff-sided lake whose waters were so dark and deep that pieces of bright, white quartz, the size of walnuts, sank slowly out of sight in the mid-day sunlight underneath the boat. Try as we might we never saw the bottom. Columbine and larkspur clung to the cliffs, above the lake, larkspur in the sun and columbine in the crevasses, and the snow shelved out across the lake on the southern shore.

Avila's interests, more than mine, had been directed primarily toward literature and the life sciences. Or perhaps it would be more accurate to say that mine, more than hers, had been directed with greater emphasis toward an understanding of what we usually think of as the inanimate substructure of the Universe at large, more toward an understanding of chemistry, physics and mathematics. And I was anxious to get back to the girl's equations. The conversation in the canoe, however, had started between Avila and the girl when Avila had pointed out the peculiarity of the flower shapes and their botanical interrelationships. (Both columbine and larkspur, as I recall, belong to the same family as the yellow buttercups, already familiar to the girl in the Berkeley hills. But she had never seen columbine or larkspur.) It was not, therefore, until we were drifting far from shore that I found a sufficient lull in their conversation to allow me to feel free to change the subject and get back to the physics of that other world.

"Is today the day," I asked, "to ask how *your people* got to E equals m?

How did *they* see that mass, or what we call matter, is just potential energy?"

"Well," she said, "you remember the question you asked Richard Feynman?"

"Yes," I said, "I certainly do. He had just given a lecture on the structure of the proton at the University of California in Berkeley, and Werner Heisenberg, who discovered the uncertainty principle by following a suggestion from Einstein, was with him that evening, and they both looked through Tumbleweed at the first quarter Moon. (Tumbleweed is a nine and a half inch telescope with a short focal length, and it's done a great deal of public service. That's why it's called Tumbleweed.) Heisenberg wasn't anxious to talk to me, but I needed to talk to Feynman. And Urania Hunter, a Sidewalk Astronomer who had made an eighteen inch telescope, was there. She knew that I was too shy to introduce myself to Feynman, so she grabbed me by the elbow and waltzed me up and introduced me to him herself, as if she had known him for years, and I got to ask my question.

"I asked him whether we could consider the rest mass of the proton as just the energy represented by its separation, in the gravitational field, from all the rest of the matter in the observable Universe.

"And he answered right away. 'If the mass of the Universe is the critical mass (that is, if the cosmological expansion rate is at the escape velocity, so that gravity won't bring it back together), it looks as though you're right.' Then he added, unasked, 'The electron is purely electrical; the proton is not.'

"That's exactly the way I had it figured out," I said. "If the proton is the canoe, the electron is the electrical outrigger. Something has to balance its electrical charge. Like the little float out at the side that keeps the outrigger canoe from tipping over."

"Well," said the girl, "you remember that Ernst Mach thought that inertia here must be related to inertia in the rest of the Universe, but he didn't know how. He didn't know why. He didn't see that E equals m. He didn't see that inertia is just energy, and that things are energetic by being separated from each other against gravity. That's really why you asked Feynman."

Richard P. Feynman



"Yes." I said, "Einstein apparently thought that the inertia of a body (its resistance to changes in its state of motion) was due to its *proximity* to other massive bodies, rather than to its *separation* from them in the gravitational field, *as I had thought.* I had already assumed that everything in the Universe is wound up to some five hundred atom bombs per pound against gravity by their *separation* from all the rest of the matter in the observable Universe. I just wanted to be sure that Feynman saw it the same way.

"I had read somewhere that the splash of a ten gram marshmallow on a neutron star, with the density of a hundred thousand battleships in a one pint jar, would be a one gram splash. It would release the energy of an atomic bomb. The rest I figured out for myself. If you dropped it to a small black hole, you'd get three times that much out. And if you dropped the marshmallow to a black hole with all the rest of the matter in the observable

Universe inside, it would be a ten gram splash."

Then I turned to Avila and said, "A ten gram marshmallow is the energy of ten atomic bombs. I know, it doesn't look like that, and they'll sell you a whole bag of them at the grocery store for a dollar nineteen. They have no *idea* what they're doing."

"And why," Avila asked me, "should the electrical energy be equal to the gravitational energy? That was the question you had as a freshman when Professor Lawrence told you that the mass of the electron was just the energy required to make it that small against its self-repulsive electrical charge."

"Yes, as I recall, Lawrence put it this way. If you push two electrons together, they weigh more together than they weighed apart because the energy of pushing them together is still in there. Then he said that if you made a big electron smaller, you'd be pushing negative charge toward negative charge and making it more massive, and the work required to make the charge of one electron as small as one electron is, is its mass."

"You mean there's nothing else in there," said Avila.

"Yes, but what I saw was that spacing out against gravity and spacing in against electricity are two sides of the same thing. You can't break a cookie into larger and larger parts. I've said it in class. Spacing out *requires* spacing in."

"So *that's* how you do it," Avila said.

"Well," said the Little One, leaning toward Avila, "he asked me how *our* physicists got to E equals m. It's because they see energy as the Underlying Existence showing in space and time, and it shows in space as electricity *and* gravity, and it shows in time as inertia. In your physics books electrical and gravitational energies are referred to as energies of position in space. If you like, you may think of inertia as energy of position in time. I know, it sounds a bit philosophical, but science is just philosophy after the measurements have been made. First you guess, and then you measure."

Smiling, she said to Avila, "We *all* see a Universe spaced out in space *and* spaced out in time. We see the Sun and the Moon as they were a while ago, not as they are when we see them. And just as J.D. saw that spaced out against gravity must be the same thing as spaced in against electricity, *we* see that spaced out in space must be the same thing as spaced back in time.

Then, since the Underlying Existence shows through in time as inertia or mass, and in space as energy, mass and energy must be the same thing."

Smiling still, she turned to me and said that it was clever of Mileva to get it from the measurements, and added that she didn't know how those old physicists in India got it, but they did.

Her smiles and her laughter were enough to cheer anyone up, and her education must have been a great deal better than mine.

After testing the temperature of the lake with her hand, Avila wanted to know, if mass is simply energy, why the Little One had said that energy was the Underlying Existence showing through in time and space, and *what*, Avila wondered, was showing through.

The girl suggested that we save that for another time, but our curiosity was on the 'alert.'

When we were lying in the sun on the warm rocks, after a cold swim in the lake and a tour of the flowers on the cliff, I told them that I had learned to swim butterfly by watching Mark Spitz and John Ferris when they were ten or twelve years old.

"Where did you know them?" Avila asked.

"When I was in the monastery," I said, "in Sacramento. There was only a fence between us and the Arden Hills Swimming and Tennis Club where the Olympic swimming team was trained. And one morning when I was working on our south wall, looking down on those swimming pools, I saw those two kids swimming back and forth butterfly. I was forty-five years old and I had had a life saving certificate when I was sixteen, but I had never seen anyone do the butterfly.

"A few years later one of the brothers in the monastery, Shanta Chaitanya, had a back problem and the doctor told him to swim. Well, he had never been swimming, so I volunteered to go as his life guard, and we got to swim next door. That's how I met all those kids. In those days I could swim a mile butterfly, that's seventy-two laps, but I learned it by watching Mark Spitz and his friend who, by the way, told me that no one swims a mile butterfly.

"Sherm Chavor, the boss, was the Olympic swimming coach in sixty-eight and had the whole team on vitamin C and vitamin E at my request.

That was the year that Mark Spitz won all those gold medals."

"And how did you know about vitamin C and vitamin E?" asked the Little One.

"From Adelle Davis's little book *Let's Eat Right to Keep Fit*. Someone had a copy of it in the monastery before I left San Francisco. Vitamins C and E are anti-oxidants. They keep your breathing from oxidizing the wrong things."

After a potty break and a snack at the cabin, we got back in the canoe and paddled again to the warm rocks. But I wanted to get back to the girl's physics, so, lying on *our* rocks and wondering what *their* rocks were like, I asked the girl from that other world, "What's next?"

She laughed and said, "Geometry, or I can't answer Avila's question about what's showing through, because whatever shows through shows through in the *geometry*. It shows through in space and time."

"I always loved geometry." I said.

Then she asked me if I remembered Einstein's *four dimensional* geometry in which space and time enter as a pair of opposites. I said that I did, and she asked whether I saw that space and time appear by contrast to each other.

"*Just a minute!*" said Avila. "What do you *mean* - space and time are opposites, and they appear by contrast to each other?"

"What I mean," she said, "is that just as up is the opposite of down in the gravitational field, and just as north is the opposite of south in the magnetic field, and just as plus is the opposite of minus in the electrical field, just so time is the opposite of space in the geometry of this world, and they appear by contrast to each other. You see an event away from you in space *by seeing it back in time.* And you see everything as it was a while ago, not as it is now, and the total separation of your perception from the event perceived is always zero because if a light beam can get across from one event to the other then the space and time separations between the two events are equal and the total space-time separation is, therefore, zero.

"Einstein knew that distances in space and intervals of time are *not* objective, that is, not seen alike by everyone, and he was looking for what *is* objective.

"Suppose two space ships are approaching each other at very high speed, the passengers in both ships will see the clocks on the other ship going too fast. But after the ships have passed each other the passengers will see that the other clocks are going more slowly. That's because lengths of time, like hours and minutes, are not objective. How long it takes a clock to go from eight to nine, as seen by you, depends on how fast the clock is coming toward you or going away.

"We live in an *observational* Universe, not an *actual* Universe. That's why Michelson and Morley couldn't measure the speed of the Earth through space. In an observational Universe there is no such thing as motion with respect to space.

"So Einstein had to put time into Pythagoras's old equation to get the *objective space-time separation* between events. Between the events here-now and there-then, in his equation, the space interval between here and there comes in squared, with a plus sign, but the time interval between now and then comes in squared with a *minus* sign. So we have to *subtract* one from the other. And, if the event is something you can see, then the space and time intervals between you and what you see are equal, and there *is* no *total separation.*

"Distances in space are not objective, nor are lengths of time. It's only the *space-time* separations that are objective.

"Suppose you see an event, say an explosion, at a distance of one light year. Then you'll see it as it happened one year ago. So the distance between the place of that explosion and the place where you see it is one light year. And the time interval between the time when it happened and the time when you *see* it happen is one year. Then, since in the real world a light year is equal to a year, and since in Einstein's equation the time must be subtracted from the distance, the total goes to zero if the distance and the time are *equal.* And Einstein's not responsible for this. He was just trying to find out how the Universe is put together, and he noticed that the physicists had misunderstood the geometry."

$$S^2 = x^2 - t^2$$

Einstein's 1905 geometry

"That space and time are related in this peculiar way," said Avila, "is probably the most outlandish thing I've ever heard. But since it's been in the equations ever since nineteen hundred and five, and since no one has yet found a measurement to contradict it, it might, just possibly, be true."

In the warm sun, the girl continued. "When J.D. was at the University they measured time intervals in seconds and space intervals in centimeters. But to put them both in the same equation they had to know how many centimeters were equal to a second. That ratio, what is known in the trade as the speed of light, is about thirty billion centimeters to a second. They should have had a unit of time to match the centimeter."

"That's interesting," I said, "When I had just graduated from the University of California, as a chemist, and was working in Gilman Hall, in nineteen forty three, I remember hearing that someone, maybe Professor G.

N. Lewis, or the low temperature man, defined a 'jiffy' as the length of time that it takes light to travel one centimeter. Eldridge, in his charming book *The Physical Basis of Things,* called it an 'einstein' with a small e."

"I knew this would be difficult," the girl continued, with one foot in the water, "but let's take it step by step. First you need to see that what you call the speed of light is not really the speed of anything at all. It's just the ratio of space to time. It's simply a statement that a certain amount of distance is equal to a certain amount of time. A light year is equal to a year."

"You can't get off that easily," Avila complained, "I *don't* see it."

"Well," said the girl, with her girlish smile, "Thirty billion centimeters per second is a ratio. It's just how much space is equal to how much time, and it's not the speed of anything. But it's known in the trade as the 'speed of light' because the scientists didn't understand the four dimensional geometry of the real world." And she sadly shook her head.

"Then, in any system," Avila said, "the speed of light in that system designates the ratio of the unit of distance, in that system, to the unit of time. Now I'm beginning to get it," she said. "If time came in squared with a *plus* sign it would just be another dimension of space, and we'd still be talking about distance instead of separation."

"Nicely put," I said. And I wondered about the world where our little friend had learned all this. It was perfectly clear that she hadn't learned it in Avila's library. And I wondered how they saw things where she used to live.

I was aware, of course, of why the girl had insisted that time comes in *squared* with a *minus* sign. She was telling us that this is no longer Euclid's geometry, where everything squared comes in with a plus sign. And what she was saying was that in Einstein's *four* dimensional geometry the time interval must be *subtracted* from the space interval to get the space-time separation between two events, and that only the *space-time separation* is objective. Only the *space-time* separation is seen the same by everyone.

Addressing Avila, the girl continued, "We see an event away from us in space *by* seeing it back in time in just such a way that the total separation between the perception and the event perceived is zero. The four dimensional separation between the perceiver and the perceived is zero."

"You've certainly given me something to think about this time," said Avila. And the girl went on.

"In order to simplify the Pythagorean Theorem for two and three dimensions of space, *as you did it at school*, we agree to measure *distances* in all directions in the same units, say centimeters. Now if we want to keep things simple, when we introduce *time* into the same equations, we must find out what unit to use for measuring time so that it will correspond to the centimeter, which we have chosen for measuring distance.

"With the einstein or the jiffy, and the centimeter," she said, "the equations remain simple. Einstein must have known this."

"He did." I interrupted, "His original energy equation was E equals m. Then he saw that if we choose to measure space in centimeters and time in seconds, and if we choose to measure mass in grams and energy in ergs, we must put in the c squared to clean up the units in the physics lab. Well, he didn't mention the physics lab."

"Why didn't you tell me all that," she asked, "when I asked you what it meant?"

"Because I didn't know what you were driving at." I answered.

"Well," said the girl, nodding sideways at me, "you astronomers are more logical. Having noticed that the Universe is fairly large, you measure it in light years, and having noticed that it's fairly old, you measure it in years. Then, of course, the speed of light is one light year per year and you're okay. You see something a light year away by seeing it a year ago, and the separation between you and what you see is zero."

Later, in the canoe, Avila pointed out that space and time *certainly don't look like opposites*, and she came back with a question for the Little One. "Why did you say that space and time were opposites like positive and negative electrical charges?"

"*Are* opposites," the girl corrected with a laugh. "We say that the money you put into the bank and the money you take out are opposites because, if the two are equal, the balance remains unchanged. We say that positive and negative electrical charges are opposites because in any system where the two are equal, the total charge is zero. Likewise we say that space and time are opposites because between any two events whose space and time separations are equal, the total *separation* is zero. It's objectively zero. It's zero as seen by anybody.

"We see the Universe spread out before us with zero separation between us and what affects us by gravity, and with zero separation between us and what we see."

I had heard a good many lectures on relativity, and I'd heard Heisenberg, Feynman and Bohr, and I had read a good deal, but I had never heard it put quite that way. I had never heard it mentioned that space and time are opposites. And I knew that Einstein had not seen it that way.

"Now I understand, at last," said Avila, "why you laid such emphasis on the fact that time comes in squared with a *minus* sign. It's because in the Pythagorean Theorem for two dimensions of space, the hypotenuse goes to zero only if both sides of the right triangle go to zero, whereas, in the equation for time and one dimension of space, since space and time are *opposites*, the space-time separation goes to zero if the space and time intervals are *equal.*"

"Right!" said the girl, obviously pleased, "But it's not a trivial case. It covers every event of your waking life. If the message is electromagnetic or gravitational, the separation between the emission event and the event of perception is zero. Whatever you see, you see in the past. You cannot see anything when it happens. You see an event away from you in space *by* seeing it in the past, and the farther away, the longer ago. If, here and now, you see the explosion of a star eight and a half light years away, then you'll see it also eight and a half years ago. If the message is carried by what you call radiation or by gravity, then the separation between the event of origin and the event of receiving that message is always zero."

"Why do you say, 'what *you call* radiation'?" Avila asked.

"Because photons don't exist." the girl answered, "There's no separation between their emission and absorption events, so they don't exist. Like field theory, they were just invented by physicists who didn't understand the geometry.

"In eighteen hundred and two Thomas Young did the double slit experiment and invented *wave theory* to explain what he saw. And in eighteen thirty-one Michael Faraday discovered electromagnetic induction and invented *field theory* to explain what *he* saw.

"You remember the problem which was connected with the famous measurement made by Michelson and Morley. Their question was asked from the standpoint of what you refer to as classical physics. They wanted to know how fast the Earth was moving through space, through what was then called the 'luminiferous ether', so they set up their apparatus to measure the speed of the Earth with respect to the 'speed of light' through that space."

She was always careful to explain things to Avila, even when she thought that I might have understood.

"Suppose we wanted to know the speed of this canoe across the water. There are no landmarks out here. So we drop pebbles into the water and measure the speed of the canoe with respect to the speed of the ripples."

"That's very clever." said Avila, "So they dropped their pebbles into the 'luminiferous ether'."

"According to classical physics that was perfectly reasonable." The girl answered. "But it's by no means reasonable if you have noticed that space and time are opposites and that there aren't any photons. There aren't any 'waves' in the 'luminiferous ether.'

"Classical physics," she said, "was developed from the point of view that reality is really spread out in space, and that time has nothing to do with it. It was assumed that space and time are not opposites. That was one of the mistakes which was taken for granted, and still is, I might add, but it was found that the real world could not be understood from that standpoint. The behavior of the real world does not correspond to that physics. Relativity theory arose. The physics was corrected, but people's understanding is still the same. They still assume that space and time are not opposites, in spite of the equations. People still assume that reality is really spread out in space, with no connection between the space and the time.

"When relativity theory became fairly well understood, the notion of the 'luminiferous ether' faded partly away, but the photons which swam in it retained their reality in the minds of the physicists. But relativity theory says that the separation between the emission and absorption of a single photon is zero. They are adjacent events in the four dimensional continuum of the real world. Now, if the separation between those 'two events' is zero, what's the meaning of asking the photon how it got from one to the other?"

"I've been wondering about that myself," I said, "and sometimes I have wondered how we could throw out the luminiferous ether and save the photons. It's like throwing out the bath water and saving the baby."

She laughed. "It's like throwing out the bath water and saving the *suds*."

"Are you suggesting, then," Avila asked, "that our concept of radiation, with its wavelengths of light, is part of the classical model and doesn't belong in modern physics?"

"The concept does not arise from the equations of modern physics, though it certainly arises in the minds of your physicists," the girl replied.

Then the girl asked Avila if she remembered the double slit experiment, and if she remembered the interference pattern that you get with both slits open. Avila said that she did, and that there was always this problem that if the photons go through the slits one at a time, how do they know that the other slit is open.

"First," the Little One said, "there are no photons, and the emission event at the laser gun and the absorption event at the screen are *adjacent* events in space-time. They're not really separate. But this adjacency has two components, a space component and a time component. And with both slits open the space component has both paths open. So the *probability* of finding the absorption event at a particular place in the screen is related to the fact that *both* paths are open.

"You remember Feynman's 'sum over histories.' You have to add up *all* the possible ways the thing could have happened. Well, if both slits are open, we have to sum over both paths. That's what gives the diffraction pattern which tells only the *probability* of finding the absorption event at a particular point in the screen. It doesn't say the photon was a wave, or that it was in two places at once.

"For *electrons*, although the emission and absorption events are *not quite adjacent*, we still sum over the *space component* of the assumed trajectory. For electrons, the time component is a little bigger than the space component. But still, as Niels Bohr has said, 'Electrons don't have trajectories.' We see only an emission event and an absorption event. That's all."

As we left the canoe, the Little One stood up and continued.

"If you fail to notice Einstein's four dimensional geometry, you'll have field theory and the wave-particle duality. Quantum mechanics is the observational evidence that the geometry of what is known in the trade as the real world is as Einstein had it in nineteen hundred and five, and that the separation between the emission and absorption events for both 'photons' and 'gravitons' is zero."

Later, at lunch in the cabin, Avila asked the girl, "What's the benefit of knowing all this physics? "Suppose we *do* have field theory and the wave-particle duality? Is that so bad?"

"It's not that it's bad," the girl said, "It's that it's wrong. You need to know where you stand. A fox knows all of the region round about her den, and *you* also need to know the physics of *your* world. Remember what James Burke said, 'If you don't know how you got somewhere, you don't know where you are'."

"Well, you may be right," Avila began again, "but I must say that you certainly have overhauled my view of modern physics. I always thought that the model of the Universe arising from modern physics was essentially the same as the classical model, but with some adjustments, here and there, for our lack of certainty, and for the fact that velocities could not be measured with respect to empty space. I never dreamed that relativity theory and quantum mechanics had introduced such a 'sea change' in our physics. I never took either one of them very seriously until I got into J.D.'s class. But even he didn't put it to me as strongly as you have.

"I think that when I understand what you've just said, I'll understand a great deal more than why the c squared doesn't belong in Einstein's equation, and why E equals m. Now, why did you say that energy is the Underlying Existence showing through, and what is it, please, that shows through?"

"Please," she said, "let's save that for another time." and we went back to the canoe.

She was very polite and, like a little kid explaining things to her parents, she was careful not to push us too hard. But it would yet be days before I even understood why she had trained us in physics. And it would be days before I even saw where her physics was to lead.

I was very glad that she had asked us to save that for another time, because, even though I had asked some of the questions, I needed time to ruminate on what we had already heard before hearing any more. Avila said later that it was like the first week in class when you find yourself in a course for which you don't have the prerequisites.

I don't know what model of the Universe Avila was able to fabricate from all that, but to me, the particle aspect of radiation seemed suddenly clear. If the emission and absorption of a single photon were adjacent events in the four dimensional continuum of the real world, then the particle nature of radiation seemed immediately understandable. The wave nature of radiation, so familiar to us all in the literature, must be, it would seem, a figment of the classical model, in which the zero separation of the really adjacent events is not seen as zero. There was, in this, much food for thought. And I wondered: Do we see the interference pattern in the double slit experiment only when we let a multitude of particles go through? It must be so, I thought, as Amit Goswami said. The diffraction pattern tells only the *probability* of finding the absorption event at a particular point. *It predicts only the probability.*

We paddled and drifted around the lake, and the ensuing conversation turned on Avila's question about the raising of children. Are they raised by the mother alone or with the help of a father?

"We have found," the girl said, "that children usually turn out better when raised by more than one parent, but sometimes we use more than two. Since our social relationships are not dominated by sex, as are so many relationships here, our social structure is a great deal more flexible. Probably you are largely unaware of the pervasive influence of the boy-girl relationship in your social structure. Even in the homosexual relationships the *type* persists.

"Our breeding methods are very different from yours and have given rise to great changes in our interests and pursuits. Your teenage boys, racing around on motorcycles and fast wheels to prove their manliness - I can't tell you how quaint it looks. We would class them as primitives. Our children are not born of such fathers.

"When you think of Einstein in his fifties or sixties - when you remember his persistent enthusiasm in searching and probing, seeking always for a deeper and more fundamental understanding of the Universe, seeking to comprehend the underlying nature of the real world - when you think of that, you are in a better position to understand. *Our* interpersonal relationships are based more on such interests. Friends and living companions are chosen more on similarities of approach to such problems than is at all common here. You two, of course, have such interests in common, and I am very happy to have been picked up by you rather than by those kids on fast wheels. But you must know that your case is not the rule. Most people here are not held together by a persistent effort to understand.

"Where sex relationships have been nearly rooted out, relationships like yours become more common, but often involve more than two persons, often three or even more. Children may be raised by any such group which, by common consent, is deemed suitable. Probably we lived longer in the sea than you, and we agree more easily. But the real mother, if at all possible, and usually the real father also, are members of the group. Raised by such a group, the child may feel that she or he has more than one mother or more than one father. But very often it is just like here; the child is raised by his real mother and his real father. We call them gene mother and gene father. We feel that what a child picks up from his surroundings is every bit as important as what he picks up from the gene pool."

Her remarks about the raising of a child by more than one mother, as well as her remarks about what they had come to feel was "natural" selection, gave rise, in my mind, to some problems which were, later on, to prove important to me. In what way, I wondered, would sibling relationships, and the relationship between parent and child, be altered by having more than one mother? And, in what way would the other interpersonal relationships be altered by rooting out the sex drive?

That evening, in the absence of the girl, Avila and I had a serious talk. Both of us felt that it was now necessary to quiz the girl on the nature of the real world, and both of us knew that we would have to struggle to understand. Somehow, if possible, we must find out what she knows that we don't know.

I asked Avila what model of the Universe she had come up with after the talk in the canoe. She replied that she could not come up with a model at all, but that she had come to understand that her old model must be totally wrong.

"The notion," she said, "that we see a Universe spread out before us in space, only by spreading it back in time, filled in the grave over the coffin of my old model. The grave was dug by you, months ago, but now it's pushing daisies under a new sun. Thoreau heard a different drummer, but that girl sees a different sun. It's not easy, you know, to persuade someone that her perception of the Universe is totally wrong, but there's something in her words, even more than in yours, that touches me inside, and there's someone inside who knows that she knows."

We had great fun that summer, the three of us, swimming, canoeing, even some fishing, although the girl didn't care for fishing. There was lots of hiking and lying around in the sun without clothes. It didn't appear to make a particle of difference to the girl whether or not we wore clothes, just as it makes no difference to very small children, and she never got sunburned.

There was something extraordinarily refreshing about that girl's companionship, and Avila remarked, "That girl is totally without hang-ups. I have never seen anyone so free."

"Youth is wasted on the young," I said, "because they don't have knowledge. But all the beauty of youth is hers, and all the knowledge and wisdom of old age."

"What a jewel!" said Avila, "What a priceless, well cut jewel! Wouldn't it be nice to visit her homeland."

"That," I said, "would be a home *away* from home."

Later, that night, we talked again, Avila and I, about that girl's physics. Both of us were anxious to get back to it, but both of us felt that we had homework to do before we dared ask her again.

We worked at our homework. We tried to feel, when we looked out into the depths of the night sky, that the events that we see are spaced out in the darkness by being spaced back in time. And we tried to feel that we were looking *down* into the sky, instead of *up*. We tried to sense the projection of space and time at the constant ratio of thirty billion centimeters to a second, or a hundred and eighty six thousand miles to a second, which was a little easier to handle, or at the ratio of one light year to a year, which was easier still. We tried to think of it as a ratio rather than a speed. What a strange geometry!

Gradually, as we talked, it became clearer. Now I could see that many of the things that I had said in my classes were wrong. Many of the things that I had heard about relativity and even taken for granted, were wrong. Now I could see that William Kaufman was wrong when he said at Davis, a couple of years ago, that when two observers are moving with respect to each other, and measure the separation of the same two events, where one sees more of space and less of time, the other sees more of time and less of space. I forget his exact wording, but I could see now that that would be true only if time behaved like another dimension of space, and entered the equation with a plus sign. It seemed to me now that what Kaufman should have said that night was that where the one sees more of space, *he* sees also more of time, and the *other* observer sees less of space and less of time between the same pair of events. It's the *difference, not the sum*, of the squares of the space and time separations that must be the same for both observers.

How clever of that girl to straighten me out. No one had ever explained it so clearly before - to me, I mean - and for years I had presented it wrongly to my classes. For years I had said it the way Kaufman had said it. For years I had told them that when two observers moved with respect to each other, part of what one called space, the other called time. I was just quoting what I had read in the books, not by Einstein but by those who interpret Einstein for the rest of us. Misinterpret might be a better word.

Now I could see that our misinterpretation, for I was as guilty as they, was based on our failure to notice that time is *not* another dimension of space. It's anti-space, and enters the equation with a minus sign. The *square of the time dimension* comes into that equation with a *minus* sign. Why don't the physicists see that? They graph it as another dimension of space.

The Andromeda Galaxy and companions

The night skies over that lake are extremely dark, and we found the depth of that darkness very conducive to the success of our mental gyrations on behalf of a better conceptual understanding of Einstein's insight into the geometry of the real world, but, for all that, we cut our gyrations short in the hope of waking early to show the girl the Andromeda Galaxy before morning. It is virtually impossible to see it well from Berkeley, and we had waited for this chance.

An hour or so before the dawn, Avila and I were up, and so was the galaxy, and it was beautiful. With our bare eyes we could easily see the reach of the spiral arms beyond the central bulge. We awakened the girl from sleep to show her.

That lake is far from any town, and more than a hundred miles from a town of any size, and its altitude above the sea is about six thousand feet. The night skies there are about as dark as I have seen them anywhere, yet, when we awakened the Little One and showed her the galaxy, she remarked, with conspicuous elation, "Yes, that's the way we used to see it, but through much darker skies, without your lights, without your smog, without your Moon, and without these foreground stars. How far is it from here?"

"About two and a quarter million light years," I replied.

"Maybe a little more," she said, and we went back to our sleeping bags.

# CHAPTER 4

## COME MORNING

By morning, Avila and I felt ready to probe the girl again. I refer to her simply as the girl because we never called her by any name. She didn't seem to want that. When we just referred to her as "the girl", she didn't seem to mind so much. When we called her directly, we called her by some descriptive name that came easily to mind, like "Little One" or "Gentle One", as if she were a little Indian girl. Often we called her "Little One".

Well, as I said, by morning we were ready to quiz her again. Our next opportunity came after lunch, when we were once again in the canoe. One of the nice things about being in the canoe was that there were no chores connected with being there. There would be no interruptions. So, once again, while we drifted in the sun, she began as though she already knew what I was about to ask.

"You must give up the notion," she said, "that the Andromeda Galaxy exists outside of you at a distance of two and a quarter million light years. It's the Andromeda Galaxy *as you see it now* that's two and a quarter million light years away and two and a quarter million years ago. And when you square those two numbers, and subtract one from the other, the answer is zero. You must learn to think in terms of events, there-then and here-now. And if either event can be seen from the other, the events are adjacent in space-time, and the separation between them is zero.

"You must learn to think in terms of events, and events have four dimensional addresses. If you write to the San Francisco Sidewalk

Astronomers, you address it to sixteen hundred Baker Street. Baker Street tells the postman how far east or west to walk. The sixteen hundred tells him how far north or south to walk. And the name tells the people in the building on what floor it goes. But the letter must carry a date. Even if you don't put it on, the people at the post office *will.* Four dimensional addresses are objective, and the separations between them are also objective. If you tell me you'll meet me on the sidewalk at Baker and Bush at nine o'clock on the morning of the Fourth of July, nineteen seventy-nine, that event is objective. It doesn't matter how fast the Earth is spinning or going around the Sun.

"Every four dimensional event is objective. Any mailman can find it. And the separation between any two such events is also objective. You must learn to think that every such event has zero separation from an uncountable number of other such events. The event here-now has zero separation from all events which can be seen from here-now, and also from all events from which the event here-now can, itself, be seen."

"I must say," said Avila, "that this new geometry is 'passing strange'."

"No," said the girl, "it's not. It's simply that your intuitive notions of space and time are partly wrong, and it's impossible to understand the Universe in that way.

"You must take the equations seriously," she said. "They're much more reliable than your intuition. In some things your intuition is reliable. Your intuition knows that love is true. Oneness is true. Your intuition knows that peace is true. Changelessness is true. And your intuition knows that freedom is true. Infinitude is true. But in these other things, stick to the equations!

"You see a Universe spread out, as it were, before you, only because there is zero separation between all those events which you can see, and the events of your perception.

"It's not that gravity acts at a distance, nor after a lapse of time. It's simply that the actions of gravity, like the actions of radiation, are between adjacent events."

"All this involves a 'sea change' in my understanding," Avila remarked, "which I have not yet built into my perception, but I understand, now, what you are saying, and if I do, he does. But I still don't know why you said that energy is the Underlying Existence showing through. And *what*, pray tell, is showing through?"

"Well," she said, "your *own* old physicists saw *that* one, they said the whole Universe is made of energy, and it's all in J.D.'s books. Those old physicists built their physics into their language and left it there for all to see. The Sanskrit word for this Universe is *Jagat*, the changing. But since change can be seen only in time, and only with respect to the changeless, those old physicists saw that there *must* be something *underlying* this changing world that's not in time and space, and, therefore, neither changing, finite nor divided. Their problem then was: How do we get from the changeless to the changing without changing it? And they said that it can only be by mistake, because you *cannot* change the changeless.

"So they studied mistakes. That was their great contribution. If they hadn't studied mistakes, they might never have seen it. But they *did* study mistakes, and they *did* see it. And they said that in order to mistake a rope for a snake in the dusk, there are *three* things that one must do. First, one must fail to recognize that it's a rope. That they called the *veiling* power of the mistake. Then, one must jump to the wrong conclusion, that it's a snake. That they called the *projecting* power of the mistake. And, finally, in the first place, one must have seen the length and diameter of the rope, in the partial light of twilight, or one never would have mistaken it for the length and diameter of a snake. That they called the *revealing* power of the mistake. Everything that happens here happens because of that revealing power, because the Underlying Existence shows through. You can't mistake your friend for a ghost without seeing your friend, because your friend is in the ghost.

"The *old* words for these three powers were red, white and black. Black was for the darkness of dusk. White was for the partial light of twilight. And red was for the fact that you *colored* it with your own imagination.

"*Our* old physicists saw things in much the same way, but you have people here who have seen *through* the mistake, saints who have realized the Underlying Existence. You have Buddha, Jesus, and the Old Man in J.D.'s shrine (She always referred to Sri Ramakrishna as the 'Old Man in J.D.'s shrine.'), and so many others whom I could name. And *that*, to me, is new.

"You remember that I said that your intuition knows that love is true? That's the undivided showing through. Freedom is the infinite. Peace is the changeless. Everything that happens here happens because that Underlying Existence shows through. The saints run after it. The genes run after it. The hydrogen runs after it. But only the saints know how to reach it. And, to me, *that's new*.

"And if seeing it thus weren't due to a mistake, faith would play no part in spiritual practice. If your house has actually burned down, faith that it has not will get you nowhere. But if it's just a mistake, there's a way out.

"The Universe isn't made out of forces, momentum, angular momentum or electrical charge. Their totals go to zero. It's made out of energy, and energy is that Underlying Existence showing through in time and space. The Changeless shows through in the changes as inertia. The Infinite shows through in the teeny weenies as the electrical energy of the minuscule particles. And the Undivided shows through in the dispersion of the particles through space as gravity, and also as the attraction between plus and minus and spin up and spin down. *Therefore, all this*.

"Those old Indian physicists, of course, didn't know about hydrogen and all that. And your modern physicists still don't see that gravity, electricity and inertia are the Underlying Existence showing through.

"From reading your astronomy books, one can readily see that you now understand that the Universe is made, by transformational causation, from what J.D. calls the primordial hydrogen. The things which you see are simply transformations of this hydrogen by gravity and so forth."

"Yes," I said, "that's gradually become clear since the advent of a little book called *The Nature of the Universe*, by Sir Fred Hoyle, in about nineteen fifty. I don't think it's in our library."

"No," she said, "it's not. But, book or no book, that represents a very fundamental step in your understanding. Once you understand that all this that you see is a transformation of hydrogen, and once you understand the laws which govern such transformations, then your problem becomes very much simplified. Once you see that, given the hydrogen and its gravity, all this will follow as a matter of course, then you see that if you can understand hydrogen you can understand anything. Then your problem becomes: Whence this hydrogen?

"But hydrogen does not arise by transformational causation from something else. It arises by apparition, by mistake, by what your old physicists called *Maya*, and it recycles from the border of the observational Universe by what your modern physicists call tunneling.

"Hold on!" cried Avila, "What do you mean, it recycles from the border? How does it recycle?"

"Well," she said, "Out there near the border, as seen by us, all these things approach zero, the energy of the radiation, the energy of the particles, their mass and their momentum. But if the momentum approaches zero, so does our uncertainty in that momentum. And therefore, by Heisenberg's uncertainty principle, our uncertainty in where they are must approach totality."

"You mean they can appear anywhere, by magic?" asked Avila.

"Yes," she answered, "*anywhere*.

"*Our* physicists, as you would call them, never came up with what Sir Fred Hoyle derisively called the Big Bang. They had a recycling model through what you call Heisenberg's uncertainty principle. And they knew that the Universe is made of frustration or it couldn't go on like this.

"In the Big Bang model the Universe, like a wound watch, has been running down toward 'heat death' for several billion years. Only in a recycling model does the Universe stay wound up. And only if it's made of frustration could it go on like this.

"The stream isn't happy when it reaches the sea. The sea, in its attempt to reach the center of the Earth, is frustrated by the rocks and the iron of the Earth's core. And the Earth is frustrated by its inertia. It can't fall into the Sun. And the Sun can't fall into the center of the Galaxy because of *its* inertia. And because of the cosmological expansion, the Galaxy can't merge with the rest of the matter in the observable Universe. And, because of the recycling from the border, the density of the Universe doesn't go down, and its entropy doesn't go up. The negative entropy for everything that happens in this Universe must be recycled, with the hydrogen, from that border.

"Don't you see? It's made of frustration because it's a mistake, and you're restricted to the '*pursuit* of happiness' not to its attainment.

"Had *your* old physicists known about hydrogen, they would have said that it arises by *Vivarta*, by mistake, and that the rest of the Universe arises *from* that hydrogen by *Parinama*, transformational causation."

"So that's how it goes." said Avila, "Everything in the physical Universe runs after the changeless, the infinite, the undivided. That's what shows through. That's where we get our physics. And everything in the biological world runs after that same Underlying Existence, and that's where we get the drives of our genetic programming. Thank you! Thank you! And I suppose the microwave background radiation, which the Big Bang people take as the proof of their model, is just the radiation that gets thermalized near the border by running through a field of low mass particles. Thank you again!"

"But you must remember," said the Little One, "that this is all *within* the mistake, and that your genetic programming won't get you out, nor will your physics. You must undo the mistake. That's what your saints discovered. They undid the mistake. For them, *that's* the name of the game."

"I wish I had known all this," I said, "when I joined the monastery long ago. I was never able to make the connection between the physics and what we were trying to do. We were trying to see through it, and I could see that the beauty in reading the lives of saints is that you can see: *Lordy, Lordy, Some one succeeded. Some one actually did it. Some one actually saw through this to the end.*

"Something of that must have crossed my mind when I first heard Swami Ashokananda speak. I was an atheist in those days, and the composer, Lou Harrison, took me to a lecture. It was Sunday, at the Century Club in San Francisco, in February of nineteen thirty-seven. I remember it clearly. I was a confirmed atheist, a belligerent atheist, going to a Sunday lecture. But when Swami opened his mouth, I knew that I had made a mistake. Either there's something behind this that I haven't noticed or this man's not here. It's like what Einstein noticed as a boy when someone gave him a magnet. As he said about it later on, 'Something deeply hidden had to be behind things.' That's how I saw it. Something deeply hidden has to be behind this world.

"It wasn't till nineteen forty-four that Swami let me join the monastery, but by then I'd heard hundreds of lectures and was persuaded not only that there's something behind all this, but that there's something to do about it."

"What *did* you do in the monastery?" Avila asked.

"I was the *pujari* in the monastery for twelve years." I replied. "I did the

*Puja*, the morning worship. We were up before five, showered and into the shrine to meditate for an hour or so. But sometimes, even before going into the shrine, I would stand by the radiator for a long time worrying about why the Universe is the way it is, and what it is that we were trying to do.

"After the meditation I cleaned the shrines, picked the flowers for the worship, and changed my clothes for class. That class was the most strenuous ordeal that I've ever been through. We were to have memorized the text for that morning, and Swami would ask us what it meant. Why does it say even? Why does it say but? And we were hopelessly bad at it, and it was *very* trying. Once I asked *him* a question. He flared up and made *me* answer it. Those were trying times and the diet was very poor. I was under a doctor's care for a B vitamin deficiency.

"We knew that we were short of B vitamins, but someone had told Swami that yeast causes intestinal ferment, and someone else had told him that wheat germ causes some other problem, so yeast and wheat germ were forbidden, and we were on vegetarian diet with no good source of B vitamins.

"I had the text book symptoms of vitamin B3 deficiency, alternate constipation and diarrhea, and being terrified with or without cause. If I heard a footstep on the stairs, or the doorbell, or the phone, my heart would pound, bang, bang, bang. I knew enough about physiology to know that adrenaline was being pumped into my blood without cause, and that it was toxic to the nerve tissue. I also knew that the way you use up the adrenaline is by heavy use of the large muscles, so if possible, I would go for a run. But it wasn't always possible. And it wasn't till one of the brothers had a disc operation and the doctor recommended yeast for the recovery of his nerve tissue that we had yeast in the monastery.

"That man said, when we put him on yeast, 'It's as if I've spent my entire life in the fog, and this is the first time I've ever seen out.'

"And for years I'd been unable to meditate for more than a few minutes at a stretch before my heart would pound. So when we had yeast in the monastery, I took some myself. Then I found that all I had to do was to sit down on the floor. 'It meditates by itself.' Then my dietary problem disappeared and I escaped the doctor's clutches.

"Before the morning class, while I was cleaning the shrines and picking flowers, the other brothers had breakfast. After class they studied and I did the worship till about noon. Then there was lunch to get, house cleaning, gardening and shopping to do, then dinner and the evening reading class. Then I washed Swami's dishes upstairs and went to bed after eleven. And I never got enough sleep. I could fall asleep standing up.

"That's sad," said Avila, and the Little One agreed.

"Yes," I said, "That's sad, and I think I could have accomplished much more if I hadn't been so sleepy.

"On Tuesdays Swami took me to the Berkeley Center to work in the garden there while he conducted interviews. Mrs. Messersmith was in charge of the garden there and introduced me to a little book called *The Chemistry of the Garden*, by Cousins. Cousins was not a chemist but he knew what to do. You put in the plant foods in liquid form but in such an order that they get precipitated in the soil. So I had to clean up his chemistry, and then it works like gang busters. We had flowers like we never saw before. Once I tried to kill a pelargonium by *over*-feeding it that way, but it went gloriously to growth and flower."

"I knew you knew gardening." Avila interjected. "We used to garden together after you got out. What else happened?"

"Well," I said, "for the last three years of my stay in San Francisco, a Canadian physicist named Michael Fell stayed with us and he and I talked physics and cosmology, almost continuously, for three years. At Swami's request I had fairly recently graduated from the University of California as a chemist, and Swami assigned to the two of us, Michael and me, the job of putting science and Vedanta together. But Michael couldn't stand Advaita Vedanta (non-dualistic Vedanta) so we had a *lot* of trouble.

"During Swami's lectures in the auditorium, if Swami was talking Advaita, Michael would fall asleep. Once he even snored in his seat. If Swami was talking dualism or Buddhism or anything else, Michael would sit bolt upright, but if Swami got off on Advaita, Michael would fall asleep. It was very funny. And, as you very well know, I'm an Advaitin, and Michael couldn't stand it. There was only one of my ideas that he liked, that magnetism might be the Coriolis Effect of electricity. I'll explain.

"Swami Brahmamayananda had asked me if we could divide energies into two categories, condensational and dispersional. Well, I had just come from the University and I said, 'No Swami, we *never* do it like that.' I was very rude. But he showed me where he had underlined in red, in Swami Vivekananda's *Complete Works,* that 'everything tends toward infinite condensation,' and, in another place, that 'everything tends toward infinite dispersion.' Gradually, after a great deal of thinking, I noticed that gravity tends only to condense nucleons, bringing them together, and that positive electrical charge tends only to disperse them. Maybe gravity is condensational and electricity is dispersional. Do we have any example of a dispersional gravitational field? We do. It's the centrifuge. It's only two dimensional but it has a Coriolis Effect. (That's what throws you down sideways when you walk across the merry-go-round.) Maybe magnetism is the Coriolis Effect of the three dimensionally dispersional electrical field. Like the Coriolis Effect, the magnetic field acts in a direction perpendicular to the direction of motion, and in proportion to the speed. Very strange, I thought.

"Well, Michael liked that idea, and he was sure it had been worked on before. So he promised to look it up when he went to Cal Tech. Later on he told me that there was no sign of it at Cal Tech.

"Anyway, after two years, I wrote a paper on what I thought about the connection between physics and Vedanta, and I offered it to Swami to read. He refused to read it on the grounds that he didn't have enough scientific background. But Michael was smarter than I was; he put *his* paper on the stair post and Swami apparently read it, because for the next several months, every time he saw me, he scolded me for some notion that had never crossed my mind."

The girl asked if messages from our floor to Swami's floor went on the stair post. I told her they did, and continued.

"I never objected to being scolded, you know. After all, one of the things we try to do in the monastery is to squelch the ego, and being scolded with or without cause is helpful, especially without cause, and especially if you don't react.

"It was only years later that I figured out what had happened. Swami must have read Michael's paper and thought that either I, or both of us, had written it. And he must have thought that I had thrown over all his teachings on Advaita Vedanta. That's sad. That hurts. I think he went to his grave thinking that.

"I sometimes think that I would still be in the monastery except for that misunderstanding. I was asked to leave because I was *reported* missing. But I *wasn't* missing. I was weeding between the sidewalk and the front wall with Mrs. Vitt and her two kids. If I had been tried, even if my accusers had picked the jury, I wouldn't have been convicted.

"Long after Swami died I used to dream that he's still alive and that he's let me back in the monastery, *and there's nothing in this Universe that can get me out again.*

"Those were heavenly dreams, but gradually I came to realize that I had to be out of the monastery to get that job done, to put science and Vedanta together. I could never have done it if I hadn't had to teach all those cosmology classes in the adult schools. And I see now that I also had to meet the two of you, first you," I said, nodding to Avila, "and then you." I said, pointing to the Little One, and she smiled and said, "That's polite, and I'm very glad you found me in that snow bank, otherwise I might still be there."

Then the girl asked me why I had never told her all this before, and I said that it was because the subject had never come up. And I said that I'm not clever enough always to understand what *does* come up. And I said, "Swami told me an interesting thing once, which I didn't get at first. He said, 'It's not sufficient to realize the truth. One must also understand.'

"At first I thought that he had said it wrongly, and that what he had meant to say was that it's not sufficient to understand the truth only in the intellect. One must also realize it in one's heart, in one's life. But later on I thought, 'No, he couldn't have made that mistake, he used to be an English teacher. He must have meant it the way he had said it.' But I wondered why he had put it *that* way.

"Later, I saw what he had meant. Many religious people have some sort of spiritual experience, but they don't *understand* it. They lack the philosophical background against which they *could* understand their experiences, and they become fanatics.

"Fanaticism arises through a lack of understanding of your own position, and a lack of appreciation of the other person's position.

"In Medieval Europe, we had the Spanish Inquisition. The religious people couldn't understand the scientists, and the scientists couldn't understand the religious people. The religious people couldn't even understand each other. They even burned their own people. And how many people had to die for that!

"And we had the Crusades and the religious wars of the Muslims, and the burning of witches in Europe and America. And even now the scientists and the religionists don't get along because, as you say, the scientists see things from the top and the religionists see things from the bottom. But if we all saw things as *you* see them, all this trouble would be gone.

"Both science and religion arose at the beach when we postponed the maturation of the brain. Remember *The Descent of Woman* by Elaine Morgan? When we prolonged our youth at the beach, we extended the curiosity of childhood into our adult state, and that is what gave rise to both science and religion. All of us are trying to figure out what's going on here. But we don't need to fight over it.

"Wouldn't it be nice to have a movie about what would be left in our physics if you could get rid of ideas by torturing people or by burning them at the stake."

By then the Sun was over the hill and the down-hill wind had come. We grabbed our clothes and paddled for the dock.

That night, at dinner, something came up for which, as usual, I was not prepared. I had noticed, of course, that the girl's taste in food did not correspond well with either of ours. Occasionally she seemed surprised when she started on the dessert. Neither of us had noticed which desserts seemed to surprise her, but this night we were about to find out.

Avila, at the prompting of some childhood memory, had fixed up a bread pudding, one of those things you make with stale bread, all sticky with something sweet. When the girl started on the pudding she looked almost as enthusiastic as someone eating a candle. Quickly my mind flashed back over the amino acid problem, and the sugar problem, which had crossed my mind more than a year ago, when we had first suspected that her biological ancestry might be different from our own. (Proteins can be made of either right handed or left handed amino acids, and the living organisms on this planet use only left handed amino acids. If her body had been made of right handed amino acids, although she might look like us, she couldn't survive on our food.)

The 'chicken soup' problem had been refreshed in my mind in connection with the life-detecting experiments on Mars, the year before, and I had enquired, at that time, of a lecturer in Berkeley, "Did we offer the Martian beasties right or left handed 'chicken soup'?" "Both," was his reply. We had offered a carefully prepared broth of both right and left handed amino acids; so that all we knew was that they 'seemed to be eating.' We didn't yet know whether 'they' were chemicals or beasties. Another time we'd offer, separately, first the one and then the other, in an effort to find out whether, if they were beasties, their biological ancestry was different from our own, or might possibly be the same. If they 'ate' only the right handed amino acids then their diseases could not infect us nor could ours infect them, and even if we ate 'Martian trout' it would not, for us, be food.

When first we picked the girl up in the snow, we had considered the possibility that she might be a political refugee, seeking asylum from some foreign land. It did not, at first, cross our minds that she might be 'foreign' in a different sense, that she might be 'biologically foreign.' And by the time the problem arose in our minds, she had, days ago, eaten our food and filled out. The inquiry, then, was needless.

Seeing her eating the pudding, my mind flashed back over the problem, and I asked her, "What's the matter, is it made of the wrong kind of chemicals?"

Avila had been busy at the stove when the girl had tried her first bite of the pudding, but was back to the table by the time I phrased my question.

"No," the girl laughed, as if she already understood the problem behind what I'd asked, "excuse me, but it's not that. It's just that our taste buds are programmed a little differently, and it's hard to get used to eating, for dessert, as you call it, something which, to us, doesn't taste yummy."

"Like your own bodies," she continued, "our bodies are primarily 'sugar burners', as you would put it, and most probably, long ago, our taste buds were programmed to read as 'sweet,' those foods which were high in our primary food need. But starting even before we took our gene selection into our own hands, 'natural selection,' as you call it, had, to a certain extent, already reprogrammed our tongues to a far more sophisticated preference. It was probably the result of our habit of not bearing offspring till we were fairly advanced in age. That left plenty of time for the people with bad eating habits to die out before their time for child-bearing matured.

"It's just my guess, from watching you, that some such thing must have happened back home. But at any rate our tongues are not now programmed to select what you call sweets. Forgive me! I should have been more careful."

"Forgive *you*? Forgive *us*!" exclaimed Avila, "Why should we have assumed that your tastes are primitive like ours? Nothing else about you is. I'm sorry, why didn't you speak up?"

"You never asked," the girl replied, "and anyhow I'm, by now, accustomed to seeing all sorts of oddities which you could never notice."

"I presume that when we picked you up in the snow drift, you knew you weren't at home," Avila said. "Did you wonder about our food?"

"Yes," she said, "that's why I nibbled so carefully at your food, on that first day. I was wondering whether, if it contained the wrong isomers, it would taste like food at all. Probably it wouldn't, but at any rate I very soon felt the unmistakable effects of nourishment and continued to eat what you fed me. You have both been very good to me. You can see how strong my body is with all this good food and exercise.

"I have no recollection of the snow drift, you know, where you picked me up. I must have been almost totally inanimate, with both motor and sensory systems shut down, and with the vital functions almost totally suspended. The first thing of which I was aware was the *warmth* of that *fire*. It felt *indescribably* good. Nothing, since then, has ever felt quite like it.

"I have no idea how they did that to me, no idea how long I stayed like that, no idea how I got into that snow drift unbroken. All I know is how it was where I was, and how it is where I am. And I certainly owe this last part of my life to you.

"Have you ever wondered," she continued, "what it would be like to be on another planet, around some other star?"

"Yes," I said, "I have."

"That's where you *are*," she replied, and all of us laughed.

Later that night Avila said to me, "When we say changeless, infinite and undivided, we mean changeless in time, and infinite and undivided in space, but that can't be what she means."

"No," I said, "it can't."

## CHAPTER 5

## THE UPPER LAKE

Our next opportunity to probe the girl's physics came the next day while we lay in the warm sun on the smooth, bare rocks *above* the cliff, beside a smaller lake whose waters, though as cold, were not so deep. Here too, as at the lower lake, a snow bank edged the southern shore, protected from the sun by trees instead of cliffs. The waters of this upper lake drain down the cliff, in splashing rivulets, into the thickets of white azaleas which crowd with fragrance the western shore of the lower lake, below the cliffs.

Into this summer paradise we had come in search of the quiet solitude befitting the meeting of two worlds, her world and ours. Thus far the rain had fallen from her world to ours, more welcome than a shower in the desert. And we had yet to know what rain would fall the other way.

Quietly I lay there for a while, meditating, if you like, on our good fortune. Avila broke the silence, but not the stillness, when she turned to me and said, "We're here for the whole day."

"Yes," I said, "what could be better?"

Only in emergencies would the girl speak to us, unless we spoke to her, or unless she was continuing some previous conversation. She wouldn't ask. She wouldn't tell. And I marveled at her self control, that utter dependence on the way things would fall out of their own accord, what the Vedantins call 'the way of the python.' That great reptile, they say, makes no effort to seek out its food. It simply waits. Perhaps it was that trait that brought the girl here. Perhaps whoever sent her knew what might survive. I thought of a passage from the Gita, 'Truthfulness is the austerity of speech; silence, the austerity of the mind.' What world view, and of what other world, lies hidden in the silence of her mind?

That afternoon, as we lay in the warm sun, I asked Avila whether she saw Heisenberg's uncertainty principle as an evidence that seeing things in time and space was due to a mistake.

"Yes," she said, "I do, and the other one is Einstein's geometry. His separation equation puts the separation between the perceiver and the perceived at zero even if you see what you see in time and space. And Heisenberg says that if you see what you see in time and space, you can't quite tell what it is that you see. It's like not being able to identify the snake when you've mistaken the rope for a snake."

"That's funny," I said, "For years I wondered whether first there was an uncertainty, and then we saw things in time and space; or whether first we saw things in time and space, and that was what imposed the uncertainty. That was one of the problems I had when we were on tour with the telescopes through the national parks with the Sidewalk Astronomers. But when we were at Glacier Point in Yosemite I finally concluded, on one of our trips to the milk store six miles away and down four thousand feet, that the uncertainty is imposed on us *by seeing things in time and space*, and that seeing things in time and space is what those old physicists called the mistake. *That's* the *source* of the uncertainty."

The twenty-four incher at Glacier Point
The milk store is down below.

"Wondering," Avila said, "is what you do best, and I caught it from you. And now I have a question for Little One. Would you please be so kind as to summarize what you said about E equals m?"

"Certainly," she replied, "You remember what J.D. said, that spacing out *requires* spacing in, and that electricity and gravity are two sides of the same thing. Hoyle also sees it that way. Likewise spacing out in space *requires* spacing out in time. We see events away from us in space *by* seeing them back in time, yet the total separation is zero.

"Then, you remember what those old physicists said in India, long ago, that there *must* be an Underlying Existence, and that it *has* to be changeless, it *has* to be infinite, it *has* to be undivided, and that it *has* to show through. Because of the revealing power it *has* to show through. You can't mistake a rope for a snake without seeing the rope. You can't mistake your friend for a ghost without seeing your friend, and your friend shows through in the ghost.

"And you remember what *I* said, that energy is that Underlying

Existence showing through. It shows through in time as inertia, and in space as electricity and gravity. You remember that gravity and electricity are energies of position in *space*. Inertia, if you like, is energy of position in *time*. But that which shows through is *one*, not two. So even if the one shows through as two, as energy and mass, or inertia, the two must still be *one*. That's why E equals m. That's why energy and mass must be the same thing."

Avila stood up, and with the magnificent gesture of a medieval knight swinging a feathered hat across his chest, she bowed to the Little One and thanked her.

"Your remarks," said Avila, "have completely rearranged my understanding of this world, and of Einstein's equations. No one, writing shorthand, writes so much in so short a space. One could write a whole book on one of his equations. In all my university days I never understood the value of an equation as I understand it now. And I suppose that so long as the physicists hang on to their notion that Einstein's equation means that matter can be *converted* to energy, they can hang on to their materialism. That is, they can hang onto the notion that the world is made of *matter* and that only a little bit of it is converted to energy in an atomic bomb."

Then I added, "I suppose it's the changeless showing through as inertia that persuades the physicists that the world is made of matter."

Addressing the Little One, Avila said, "If I understand you aright, that seeing things in time and space is a mistake, and that energy is the Underlying Existence showing through in that mistake, then gravity, electricity and inertia do not arise *within* the domain of our physics. Rather they *cause* our physics, and that is why Feynman says that no mechanism is known for gravity or inertia."

"One of the things I always liked about Feynman," I said, "is that he's always careful to point out, along with what we think we know, what we *don't* know."

"I have it from Avila," said the Little One, "that you would never be intimidated by the fact that no one ever understood something before."

"No," I said, "I was never intimidated by that. If we're to be intimidated by that then no new knowledge would ever be known. That's one trouble with the academic community; you're judged by what's already known. It's inimical to change. The Nobel Prize Committee told Einstein, when they gave him the Nobel Prize for quantum mechanics, that 'This has nothing to do with your relativity theory.' They have to wait to see what it means to the academic community. But I have often wondered about the things that Feynman says we don't know. With what are *they* connected? So now I see it. They're connected with the Underlying Existence showing through. No *wonder* I didn't see it. *Nobody* sees it."

Then Avila continued, "So it's the genetic programming that has us running after peace and love and freedom. And, as J.D. points out, the hippies had symbols for peace and love but they didn't have one for freedom."

"Yes," I cut in, "And they went straight down the tubes. After the sixties there were steel bars on the shop doors all along Haight Street.

"I spent a lot of time on Haight Street in the sixties, listening to the hippies and to the preachers who were trying to convince them of the error of their ways. But I must say that I had to side with the hippies. And I went to a Town Hall meeting in a grade school I had attended in nineteen twenty seven. The meeting was on the possibility of legalizing marijuana, and there was a very interesting speaker, a lady who claimed that the marijuana people had gotten her off heroin. I wanted to talk to her afterwards but she would have none of it.

"I wrote a verse, you know, for a popular song in those days. It was for *Clouds,* by Joni Mitchell or Judi Collins. You remember, '*Rows and flows of angel's hair, and ice cream castles in the air*'? Well, I wrote another verse:

*Dugs and drugs and pubic hair, and hippie litter everywhere,*
*But love and peace were in the air. I've looked at drugs that way.*
*Now the scars will hardly heal.*
*The love and peace were hardly real.*
*So many things I've yet to feel, but drugs got in my way.*
*I've looked at drugs from both sides now,*
*From up and down, but still somehow,*
*It's drugs' illusions I recall. I never did find truth at all.*

"I still think it should be used on one of those TV programs like Canon, or Barnaby Jones, where they're talking about drugs."

Perhaps because of her southern upbringing, Avila always showed great poise and self control, and didn't often indulge in demonstrative gestures, but she slapped me on the back and said, "Why not?"


The girl was a keen observer and she liked to watch us swim in many different ways. Especially she liked to watch my butterfly. She said that it reminded her of their 'porpoises' which, she said, flew clear of the water at each stroke. From her description, we presumed that they must have been mammals, gone back to the sea from the primate state. They kept their arms, webbed their fingers, kept their legs, with frog-like feet, and remained negotiable on land, as well as in the water. Unlike me, she said, they use two kicks per stroke, but their knees bend "fore and aft" and, with the help of those great muscles of their arms and chests, they could move at high speed "like one of those stones you skip across the water."

Our whales held quite a fascination for her. She had seen them at Marine World, on the way to Palomar. She said that they had no whales and was appalled that we would kill them. Almost never, she said, would they kill another mammal.

As for herself, she loved to swim under water. Those lakes are much too cold for me, and, after a little very vigorous swimming, I'd stand in the sun and watch her. I'd watch them both. Avila, too, could loiter in that water. But that girl could stay under for such a length of time that Avila would loose track of where she was, and couldn't find her. The girl said that their bodies would store oxygen at the mere sight of a large expanse of water; so that, for her, it was no strain to stay under like that.

When the two of them came out to warm up in the sun, after a short swim in the upper lake, Avila lay face down on the rocks and let me rub her back. The girl watched. Sometimes I had rubbed her back like that, and sometimes she and Avila had rubbed each other's feet. The two of them were very affectionate toward each other, and Avila loved that girl so much that I wondered how her mind could still find room for me. She had been a student in one of my classes, and had fallen fond of me, noticing that my mind seemed taken up in the pursuit of a clear understanding of the real world.

But now that she had found this girl, so far ahead of me in this pursuit, I wondered if the two of us would slowly drift apart.

She soon had enough of that back rub, and wanted to get again to that idea, that curious connection between apparent opposites and their identity. She rolled over and sat up, and I could see that her mind was reaching.

"Wherefrom comes that idea," she asked, "that the existence of two things as opposites rests on their identity?"

"It comes from observation," the girl said, "but probably you never noticed it. Boys and girls are opposites only as people. Up and down are opposites only as directions, along the gravitational gradient. East and west are opposites along the direction of the Earth's spin. It's perfectly obvious that there's no such thing as the opposite of a dog, though, as dogs, the male and female exist as opposites.

"But I feel that you are probing for something deeper. Some opposites go to zero like momentum, angular momentum and electrical charge. That is, momentum to the right plus momentum to the left go to zero. They're identical in that they're both momentum, but the total goes to zero. The Universe is not made out of momentum, angular momentum, forces or electrical charge. Their totals go to zero. As J.D. says about momentum, 'The Universe is all dressed up with no place to go.' With respect to what could it have momentum?

"The Universe is made out of energy, and the total energy does *not* go to zero, and what your mind is asking is: What is that final identity which remains when all the opposites are cancelled out? Well, *within* space and time, energy remains; but if time and space are cancelled out, only the Underlying Existence remains, the changeless, the infinite, the undivided. But you must not think, because I used three words, that there are three entities. No. That which is changeless, *is* infinite, *is* undivided. And please notice; I used only negative terms. I didn't say *anything* about it."

Addressing Avila, the girl continued, "As J.D. would put it, the Exterior Decorator has a problem. If She makes two, the oneness will show through and close the twoness down. And if She makes many, the oneness will show through and close the manyness down. But if She makes both two and many, a duality *within* a plurality, then they can keep each other up. The plurality can keep the duality from collapsing, and the duality can keep the plurality from collapsing. Surely J.D. has noticed that Heisenberg's uncertainty principle prevents the collapse of the electrical duality of the electron and the proton in the hydrogen atom because the proton is gravitational and the electron is not."

"Yes," I said, "I noticed. And that was what was so interesting about Feynman's remark that, "The electron is purely electrical; the proton is not." But I also noticed that Pauli's Verbot (Pauli's exclusion principle) prevents the collapse of neutron stars because the neutrons are Fermi particles with only a half unit of spin, so that only two neutrons can sit in the same place."

"That," she said, "is the spin duality preventing the collapse of the gravitational plurality. You must have noticed that the quantum mechanical spin is a duality, *not a plurality*. It's only spin-up and spin-down. Angular momentum could have been in *any* direction."

"So that's how it goes!" Avila cut in, "*That's* why we see hydrogen. If we see things in space and time, it's *automatically* hydrogen."

"Yes," the girl replied, "It's *automatically* hydrogen. As JD learned from Professor Lawrence in nineteen thirty-four, the infinite showing through in the teeny weenies *makes* the rest mass of the electron, and as Feynman agreed long later, the undivided showing through in the dispersion *makes* the rest mass of the proton. The attraction between them is the undivided showing through in the duality. That's why we see hydrogen.

"The plurality and the duality keep each other up, and *only* the hydrogen arises by what your older physicists would have called *Vivarta,* by what you would call a mistake.

"And there's an enormous attraction between the electrons and the protons in the hydrogen atoms but the uncertainty principle keeps them apart. And they're wound up by being *apart*, as well as by being small.

"The Universe is wound up to some five hundred atom bombs per pound by being dispersed in the condensational gravitational field. That's your gravitational energy. And it's partly wound up by being small in the dispersional electrical field. That's your electrical energy. But it's *also* wound up by the *separation* of plus from minus and spin up from spin down.

"And if an electron gets close enough to a proton to form a hydrogen atom, the energy falls by some thirteen electron volts. And if *two* electrons get close enough to *four* protons to form a helium nucleus, the energy falls by nearly three quarters of one percent of the rest mass of the original hydrogen. In larger nuclei even more electrons can come in, because then you know even less about where they are. And the energy goes farther down. That's what's known in the trade as nuclear energy, and the biggest drop is from hydrogen to helium, where two electrons get to sit on four protons.

"Everything else arises *from that hydrogen* by what those old physicists called *Parinama*, transformational causation."

"Yes," I said, "and the details are in Burbidge, Burbidge, Fowler and Hoyle. When I was a kid, it was taken for granted that the chemical mix of the Universe had been given at the time of creation, if there was a creation, or had been around forever, if there was a forever. But the Big Bang people wanted to get the mix out of the Big Bang. And they tried, but they found that it couldn't be done. Then Burbidge, Burbidge, Fowler and Hoyle wrote a paper. Well, I knew all of those people, especially Doctor Margaret Burbidge whom I met at the Summer Scientific Meeting of the Astronomical Society of the Pacific in Ensenada, Mexico. The four of them wrote a paper called "Synthesis of the Elements in Stars." So, *after that*, we *knew* where the rest of the chemical elements came from. And by then we knew enough about astronomy and geology to put the rest together. But no one knows where the hydrogen comes from."

Margaret Burbidge

Fred Hoyle

"Yes." The girl continued, *"Only the hydrogen arises automatically, by Vivarta, by a mistake,* but the revealing power drives the consequent transformations because the changeless, the infinite and the undivided show through as gravity and the like. That's why the hydrogen falls together to galaxies and stars.

"Gravity drives the cosmological expansion by converting the gravitational energy to radiation which looses *its* energy to redshifting in the expansion. That drives the recycling from the border, and then the recycled hydrogen, with its negative entropy restored, drives the gravitational collapse. It all comes back as hydrogen or it couldn't go on like this. All this is by transformational causation. Even the bodies of living organisms arise by transformational causation, but the notion that one *is such a body* must, again, be a mistake."

Still addressing me, the girl continued, "Positive and negative electrical charges exist by contrast to each other in the electrical field. They're both electrical charges. But the electrical and gravitational fields, themselves, exist by contrast to each other in space. It's your condensation and dispersion. 'Gravity sucks and electricity pushes,' as you say. Squeezed down and spaced out are opposites in space.

"And, finally," she said, turning again to Avila, "space and time, as we saw earlier, are opposites in that Underlying Existence, and what you are asking about is *that*. What your mind is probing for is that ultimate identity which remains when space and time are cancelled out."

"Yes," Avila admitted, "but I couldn't have worded it so well. That's exactly what I want to know."

And I was reminded of a passage from the Brihadaranyaka Upanishad. Maitreyi, at the offer of wealth, had asked her husband, "My lord, if this whole Earth full of wealth belonged to me, should I become immortal by that?" "No," he had said, "like the life of the rich would be your life, but there is no hope of immortality by wealth." To which she had responded with another question. "What shall I do with that by which I do not become immortal? What thou knowest of immortality, tell me that."

It's a famous passage, and it came to mind when I saw Avila so eagerly questioning the girl. "That's exactly what I want to know."

"Well," said the girl, "As I said before, we can't say anything about it; that is, from the physics and from the geometry we can't say anything. And you remember that the Old Man in J.D.'s shrine couldn't say anything either, and he'd been there.

"But the great beauty of Einstein's special relativity theory lies in this, that it points out that the separation between the perceiver and the perceived goes to zero. And when *that* separation goes, only the Old Man in J.D.'s shrine knows what remains.

We ate our lunch, skipped some stones across the little lake, put on our shoes, and climbed, like Alpine botanists, over and across the faces of those cliffs. There were many flowers, pretty, pretty flowers, and speckled frogs, in the warm mud, along the edges of a shallow pond. The sky was clear; the air was warm, and the snows of Mount Shasta shown distant and serene, as we climbed down, by another way, to camp.

Perhaps it was better that we didn't know, at first, that we'd have so little time to spend together at the lake. We talked as if the rest of time was waiting. Perhaps it was as well we didn't know that only one more day remained for us to talk, drifting in the sun in that canoe. The next day was cool and blustery, and I took advantage of the bad weather to drive to the town of Mount Shasta for fresh supplies, and to make a check-up call to Berkeley. If only I hadn't touched that phone. But I did touch it, and by the time I'd hung up the receiver, I knew that we'd have only one more day together at the lake.

I thought of going down myself, and leaving the two of them there. I spoke of it to Avila, when I got back.

"No," she said, "Where you go, she goes, and where she goes, I go. There's no other way. It's between the three of us now."

That evening, after dinner, when the chores were done, the three of us together, eyeballed the sky. The Moon was still a crescent. It was Avila's suggestion that we show the girl, as best we could, the positions, in the night sky, of the galaxies with which she seemed especially familiar. Avila had the feeling that perhaps, never again, would the three of us be together under such favorable skies. We never were.

We had already shown the girl M31, the Andromeda Galaxy, a couple of nights ago, after the Moon was down, and she knew, of course, of its companions, M32 and NGC 205. We had in mind to show her the more distant galaxies, M81 and M82, where the water pours out of the Big Dipper, and M51, the Whirlpool, and M101, on either side of the Big Dipper's handle.

It's impossible, of course, for us, with our bare eyes, to see any other galaxy but M31 and the Magellanic clouds. But we knew her eyes were good. She had told us that they could see many galaxies with their bare eyes. And it was our hope that she could see Bernice's Hairclip, as the Sidewalk Astronomers call NGC 4565, that beautiful, edge-on spiral, already sinking in the west, and NGC 253, coming up much later, south of east.

The Little One was already familiar with our star patterns, the Big Dipper, the Northern Cross, Scorpio and so forth, and she knew that our central bulge lay in the direction of the Teapot, in Sagittarius. All that she knew.

When we'd shown her the positions of the galaxies near the Big Dipper, and then NGC 4565, in the west, almost setting in our local trees, she quietly said, "It's the same here as where I was. NGC 253, as you call it, will be a long time coming up over there." And she gestured with her head toward the south east.

"The spiral galaxy which you call M33 must be south and east of Andromeda and her companions, and there is a huge cluster of galaxies there (She pointed overhead toward the constellation Hercules.), and another cluster south of your Hairclip, over there, both of which we could see only with telescopes."

"Yes, exactly" I said, unprepared for what would follow.

"But there were two clusters, clusters of stars, one of them very tightly packed, which we could see easily with our bare eyes. Where are they from here?"

If her clusters were prominent in our skies, and if they were situated near each other in space, and if one of them was 'tightly packed', it could mean only one thing to me. They must be the globular clusters in Hercules, M13 and M92.

"Probably there," I said, pointing to the northern sector of the Keystone.

"How far apart are they as seen from here?"

"A few degrees" I guessed, and pointed them out.

She could see them both, with her bare eyes, and was quite happy about it. She had seen several pictures of M13, in my astronomy books, but never a picture of M92.

"Probably they are the two," she said, "but we seem a bit farther from them here. They were on our back side and you were on our front, but they appeared to be a little farther apart, as seen from there."

That night, when we were in our sleeping bags, I took the opportunity, when the girl went to the 'outhouse', to convey my thought to Avila. I grabbed her thumb, and rolling my face toward her ear, I whispered, "That girl might be a thousand years old."

"Too bad," she laughed.

And I laughed too.

Tomorrow would be our last day at the lake, and it might be my last day to talk to that girl, but strange thoughts were running through my head.

The twenty-four incher leaving Glacier Point

CHAPTER 6

BUTTERFLIES

Knowing that this would be our last day at the lake, we had turned in fairly early the night before, after showing the girl the directions, from our local point of view in the northern hemisphere, to the galaxies with which she was familiar. We were all up fairly early and had our 'quiet times.' The girl spent the hours of dawn alone. Avila meditated quite a bit, as was her custom. And I, as usual, walked alone in the early morning, reciting passages, mostly from the Upanishads, in Sanskrit or in English.

Quiet times over, Avila and the girl were talking 'girl things' when I came up from the lake after my walk. Noticing that the girl's menstrual cycle was much longer than her own Avila had suddenly remembered the Moon problem. The girl had no moon. What had set the period of her menstrual cycle?

Avila told me afterwards that the girl had said that she didn't know how it got started but that they had what we would probably call butterflies with photosynthetic wings. And these 'butterflies' had the same problem that so many insects have here. The members of the same species must reach the adult phase in the same time period. Otherwise mating couldn't take place.

It was at this point that I entered the conversation. They were talking about seasons. The girl had said that not only do they have no months, because they have no moon, but also they have no year, that is, there were no seasonal variations to mark it, because their sun was almost on their equatorial plane.

I became immediately fascinated and asked her on what observation their time reckoning was based.

"The day," she said, "but I haven't finished answering Avila's question."

She continued her explanation to Avila saying that they had no special season for flowers when the insects could get all the nectar they needed, and probably for that reason, the 'butterflies' depended partly on photosynthesis. But she said that it was a very spectacular display about every sixty days (their time), when the 'butterflies' would all be out. And she thought that many other creatures had regulated their biological cycles to *them*, to the flights of the 'butterflies.' She said that they were very pretty, with very large 'green leafy' wings but not as proliferated in variety as *ours*.

"When I first saw *your* butterflies spreading their wings out in the sun," she said, "I wondered about them. They have the same habit.

"We have a 'thing' about the butterflies," she continued, "as you have a 'thing' about spring, or about the Moon. We have a gene pool response to them just as you have to what I would call your 'older environment.' It's that response in your genes that made you bring me to this lake. You have a gene pool adjustment to this older environment, what some of you call the 'natural' environment, which gives rise to an emotional response of peace and freedom and well-being when you get out of the 'city.' The 'city' is your newer environment, and the 'country' is your older environment, and it has nothing to do with 'natural' or 'unnatural.' I notice that you two are much more careful in your use of those words than is usual.

"We feel peculiarly inspired when we see all those green leafy butterflies, and I imagine that in the time of our ancestors that was the time for 'falling in love.' (She wiggled her fingers for quotes.) To us it's most beautiful. The cycles of our minds are still connected there, as are the cycles of our bodies.

"In a sense it's like your full Moon, but to us, intensely beautiful, more like your spring. But the astronomers do not measure time by days, nor by the flights of the butterflies."

In answer to *my* question, she now turned to me.

"The astronomers designate time by measurements of distance like your jiffy or your einstein. Just as you designate the light year by a measurement of time, they designate time by a measurement of distance.

Their unit of distance is the distance to our star, our sun, and that unit of time is about nine of your minutes. I don't know how they knew that, but it's pretty old. It's divided into tenths and hundredths, and multiplied by tens and hundreds and so forth. We have ten fingers just like you.

"We have, of course, an older system based on our day and the tides. Our tide is always high at noon and midnight. Our word for morning means 'rising tide' and our word for afternoon means 'falling tide.' But our astronomers and physicists, as you would call them, refer to time by designating a corresponding distance. Your jiffy they would designate as a time centimeter. But we shorten the words. They would be more like 'inch' and 'tinch'," she said.

I asked her about the length of their day. She said it was shorter than here by about a third, she presumed, because her menstrual cycle there was sixty days, and here, about forty.

That conversation took place during the preparation and eating of breakfast, then Avila packed a lunch and we were off for the canoe.

The girl's story about the green leafy 'butterflies,' in some other world set my mind on an old train of thought. When I was a child, still feeling like a stranger in *this* world, I used to wonder: If I had the Universe to make, out of what would I make it? I have never felt that the existence of the Universe was quite logical. On the way down to the canoe I mentioned all this to the two of them and added that back then I certainly wouldn't have started with electrical particles.

"There's *no* other way," the girl cut in, in a singsongy voice, "Don't you *see* that?"

"Yes," I said, "I see it now, but I didn't see it then. And when I was ten, I took for granted that even if there hadn't been any creation, even if there hadn't been a Universe, there would still have been space and time."

"You Christians," she said. (I couldn't quite understand how that meant me, but somehow it did.) "You Christians like to think that the creation, as you call it, is something that took place as a process in time, first in the mind of the creator, and then subsequently in fact. But no such thing has taken place."

"Wait," I said, "till we get in the canoe."

I couldn't see it then, but I see now what she was doing. Although she rarely started the conversations herself, she watched like a cat and carefully picked out from among our stray remarks, according to some hidden plan of her own, those on which to jump. Unknown to us, the entire train of thought in these long dialogues was dictated by that girl, and the continuity of the train of thought which ran through them became clear to me only at the end when she was gone.

Into the boat we went, and out into the lake, away from shore, and out into the sun. We laid the paddles in and let it drift.

"Tell me," she said with a mischievous smile, "what you mean by creation and what you mean by causation."

Why, I wondered, does she always have the upper hand?

"Well," I said, "I was born and raised in the Methodist Mission in Peking, where creation was taken for granted, and it never would have crossed my mind when I was ten that the Universe might have arisen in some other way."

"It seems to me," Avila cut in, "that creation implies a creator who made something out of nothing; whereas causation simply implies that there was a process by which something was made out of something else."

"Yes," I said, "Causation implies the conservation laws, that nothing will show up in the effect that was not already there in the cause. The form may change, but not the substance. As Thomas Gold puts it, 'Things are as they are because they were as they were'."

"And the cause must precede the effect," said Avila, and added, "But even in creation the creator must precede the creation."

"Then, in either case," the girl chimed in, "It's a process in time. But no such thing has happened. Time is *part* of it. The Universe is the result of seeing the Underlying Existence *as* in time and space, and it's that Underlying Existence that shows through in the physics."

"So that's what Swami Vivekananda must have meant," I blurted out. "He said that nothing could arise by evolution that was not already put in by involution. This must be what he meant by involution. The reality is already in it. It was not put in through a process in time. Thank you!"

"But," added Avila, "evolution takes place in time, as we thought."

"Yes," said the girl, "but involution *can't*. Time and space are the *result* of that involution."

To which I added that Swami Vivekananda had said that the Universe is the Absolute seen through the screen of time, space and causation. So he must have seen what *she* sees.

Remembering how the conversation had started on the way to the canoe, I remarked to the girl that she must have found it quaint that I had thought of making the Universe *out of something.*

"Quaint," she said, "Avila's straighter than I thought, but I thought you were straighter than that."

Avila felt that it was time for a swim. We let the girl slip over the side, and she swam out of sight beneath the surface while we paddled for the dock. Her eyes could focus under water in a way quite impossible for us, and she soon became familiar with the contours of the under water ledge and the shallower parts of the bottom, and told us of things impossible for us to see. I never encouraged Avila to swim from the canoe in the middle of the lake. If anything should happen, I could never retrieve her in time from such a depth. *We* swam from the dock. But that little girl could stay under for such a length of time that I might have worried more about a seal.

Avila and I swam for a while, and then lay in the sun on the dock. The girl was still exploring under water. When she came out, we all warmed up in the sun and got back in the canoe.

We paddled across the deep, dark waters of the lake to a place underneath the cliffs, and between the sunken logs that stick up from the bottom where the water's not so deep. Avila dropped some tiny bits of chicken for the trout saying she'd rather feed them than eat them.

"How about the chicken?" I asked. "Isn't it upside down to feed dinosaurs to fish?"

"So what!" she said.

We had talked a lot about dinosaurs after reading in "Scientific American," two or three years earlier, that the birds were descended from some warm-blooded dinosaurs who, in order to keep warm, had invented feathers in the gene pool.

For years I had wondered about the problem of how bird wings had evolved. What were they good for in the intermediate stages before they were useful for flying? They must have had some 'survival value' as my dad would have said, or they wouldn't have survived at all. I had thought that probably they were first used for flying under water, as some birds do even now. Under water, even a rudimentary wing might be of some use.

Until we had read that article in "Scientific American," it had never crossed my mind that feathers might have been invented for warmth. So now, in our sleeping bags, we use feathers as they were used by the early birds, and with their original intent.

Avila had borrowed that article from me before we went to the Grand Canyon with the Sidewalk Astronomers, and these thoughts were uppermost in our minds when we visited the dinosaur tracks by the highway in Arizona. The depressions in the rocks there look like great three-toed chicken tracks about eleven inches long, and Avila had described them in the tour article which she had written for the Sidewalk Astronomers' newsletter.

But this time my remark to Avila about feeding dinosaurs to fish put me in mind of a problem for the girl from the peripheral star, and I wondered whether, 'at home,' she had ever seen a bird. I turned to her and asked, "Do you have birds?"

"No," she said, "If we had your flying dinosaurs those beautiful, beautiful butterflies would soon be gone. And many other changes would surely follow. Without the butterflies I might not feel that peculiar inspiration which I feel now and then. Our fish eat the butterflies now, and you could probably use orchids for bait, but our only flying things are like your insects with the bones on the outside."

Avila laughed. "Exoskeletal we say, but not with bones on the outside."

The girl also laughed, then apologized and said she couldn't always tell how we would say it, when we translated from one language to another.

We wanted to know more about the native animals on her world. She said that their mammalian past was somewhat similar to ours but that their reptilian past was different.

"We have no rattlesnakes or cobras," she said with a smile. "All our reptiles still have legs, and some of them are warm blooded and very good at jumping. They catch the butterflies on the wing."

"Do any of them sing?" asked Avila.

"No," she said, "They make sounds more like your insects, and some of the warm blooded ones have what you would call fur. But they all lay eggs, what J.D. would call *land* eggs, not laid in the water, and we eat their eggs. But the eggs of our fish are toxic. We leave them alone, but we can eat the fish."

"You must have zoologists as well as physicists and astronomers." Avila suggested.

"Yes," replied the girl, "as you have physicists and philosophers. Your physicists know one end and your philosophers know the other. But you must take the physicists more seriously, much more seriously than they take themselves. That is, you must take the *equations* seriously; you must be willing to follow wherever they lead, because *they're* based on observation.

"You already know from Heisenberg that it's a mistake, and you already know from Einstein that the separation between the perceiver and the perceived goes to zero when space and time are cancelled out. With what model are you left, then, when you put them together?"

I could see at once that there was no hope of getting a model of a Universe with zero separation between the perceiver and the perceived, because it's zero separation, not only between us and what we could see if nothing was in the way, but between us and anything that could affect us by gravity. I had never taken that zero so seriously, nor had Einstein himself, I remembered, and I had always thought that whereas formerly we had the classical model of the Universe, we now had an improved model and we were waiting for a breakthrough.

Suddenly it hit me. *That girl's not waiting for a breakthrough. She's already there. The equations go all the way through.* "Okay." I said, "You win. We don't get a model of a Universe. We get a model of what the Christians would call the "creation." But it's not a creation; it's what you would call a misunderstanding. And what we're seeing here is that Underlying Existence, but we're seeing it through the fog of space and time. It's like a dream."

"Bravo!" said the girl, and Avila joined in, and added, "So you two can see all this in the equations. Bravo! Indeed."

We took our final dip in the shallow water along the northeast shore, where the stream flows out of the lake. We dried in the sun on the logjam, and pulled out the canoe. Tomorrow, in the morning, we'd be gone.

Back at the cabin, the Little One asked me if I had ever noticed that some of those old hymns that I sing were written while Sri Ramakrishna was alive.

"Yes," I said, "I noticed it while I was still in the monastery in Sacramento. I used to sing those old things with that old Estes reed organ that we had in our beautiful library. And one day I noticed the dates on 'Nearer my God to Thee,' and that made me suspicious, so I looked up some of the others."

"What made you suspicious?" she asked. "Or were you just curious?"

"No," I said, "I was suspicious because of something that happened long ago. I was raised in the Methodist Mission in Peking, and we sang a lot of hymns. And when we came to California we joined the First Unitarian Church at Franklin and Geary Streets in San Francisco, and we sang and we sang. My brothers and I were in the Sunday School choir, and Mom played the pipe organ, one of the two pipe organs that 'came around the Horn' from Europe. And we used to go caroling through the streets at Christmas time. That was one of my most favorite things.

"And when we sang 'Nearer my God to Thee,' we always sang it in both threes *and* fours. Mom always played it that way."

"That's how you sing it even now," she chuckled.

Chuckling myself, I said, "Yes, that's how it's supposed to be, but if you look in the hymn books you'll find it's either in threes *or* in fours, not in threes *and* fours. And that's why I looked it up. But I became suspicious because of something that happened long ago at Asilomar."

"Where's Asilomar?" she asked.

"It's down the coast on Monterey Bay near Carmel. The Unitarians had a place down there where they held conferences. It was at one of those conferences, where we met the Trumpler family, that Doctor Mylenberg gave a series of talks. He was a student of Jewish history, and he pointed out that every time the Jewish nation had a problem, a prophet was born to take care of it. And just when the Jews were scattered abroad, Jesus was born.

"Well, it charmed the hell out of me, and I wondered about the rest of us."

"Who were the Trumplers?" the curious little girl asked.

I said, "Doctor Trumpler was an astronomer at Lick Observatory who determined the position of the center of our Milky Way with respect to the clusters of stars in the disc. Shapley had done it with respect to the globulars. Trumpler did it with respect to the disc clusters. The whole Trumpler family was Swiss, smart and charming, and my younger brother married one of the girls."

"Was that your doctor brother?" she asked.

"Yes," I said, "and they have five smart kids, fluent in French, because during World War Two my brother was in charge of the World Health Organization, Radiation Department in Geneva. He was there for nine years, so his kids know French. But none of this is suspicious."

"I'm sorry for the interruption," she said, "But I'm interested in all the things I see along the side of the road. Please go on."

"It was Doctor Mylenberg's remarks that made me wonder about India and the rest of us. When India had to absorb the Muslims, Sri Chaitanya was born, and he had followers in both camps. The holy men serve as glue between the camps. But now the problem is bigger. Now it's India and the West. Is that why Sri Ramakrishna came and brought Swamiji?"

"Do you suppose," she asked, "that the Exterior Decorator has her fingers in the pie?"

"Why not?" I said, "Whose pie is it?"

She just laughed.

"Well," I said, "I thought that if it had happened with the Jews and if it had happened with the Hindus, what about the rest of us? The problem now is between Europe and Asia, and between science and religion, with a lot of other problems thrown in. Do you suppose that's why the Old Man, as you call him, came and brought Swamiji?"

"Maybe so," she said, "but I'm delighted to be mixed up in it."

"And I, too, am delighted that you're mixed up in it," I said, "because for a long time I've been very much annoyed that the job of fitting science and religion together had been dumped on me. It should have been finished by Swami Vivekananda. But back then the science wasn't ready. So he tried to fix it. Back then we didn't have Einstein's geometry, and we didn't have E equals m. But I didn't understand all this at first. I didn't see Swamiji's problem back then.

"When I first went to Swami Ashokananda for instruction, he asked me if I got enough sleep, and if I ate properly. I told him that I slept very little and that sometimes I didn't eat. He asked me to change that. Next time I went down the aisle to shake hands with him, after a lecture in the Old Temple, he asked me how I felt, and I told him that I was eating more and sleeping more and that I felt better. Then, when I was more than half way up the aisle, he *called* out after me, 'Eat well! Sleep well! Don't tire out the horse! You've got to ride him a long distance.' Everyone who heard it remembered it. And now I see what it means, I have this job to do."

"Mother doesn't get everything done by only one person," she said, "Haven't you noticed?"

"Maybe so," I said, "But people would have listened to Swamiji."

"They listened to him while he was here; maybe they'll listen to you when you're gone."

"Who knows? But I have faith that the scientists will eventually straighten things out, even the Big Bang people. But I'd like to be in the position of the man who discovered the Missoula Floods. He was able to say, 'Those to whom I would have said, I told you so, have died.' Wouldn't that be lovely?

"Once at the Ganges Monastery in Michigan, a lady said to me, 'You're a man with a mission.' I said, 'Lady, you've *got* it.' And once I told Ruth Ballard that I thought this body was being kept alive to get this job done. Then, in nineteen seventy four, when I thought I could put science and Vedanta together, I said, 'Now that I can do it, maybe the body will be allowed to go.' 'No,' she said, 'now you have to broadcast it'."

The Little One gave me a hug, and we looked for Avila.

Then I told both of them that when I was a kid in China, maybe nine years old, I had longingly wished that I had been born soon after Jesus was here. Little did I know how things would turn out. I didn't know then about Sri Ramakrishna and Holy Mother. But I was *alive* while she was here.

Then I said to the girl, "In India they say that when a holy man is born it's like spring. I think Holy Mother said something like that about the Old

Man, and someone said, 'Then how come so many bad things happen?' She said, 'In the spring, *even the weeds* flourish.'

"But many people feel that spiritual practice is easier since the Old Man was here. And that's why I looked up the dates on those hymns. How come I'm so moved by the hymns that date from his lifetime? And 'Thou art my All in All' and 'Thou hast lifted all my sorrow' are from M's Gospel. That's why I was suspicious. Those are the songs the Old Man used to sing.

"And he said that he would come again. I think he said in a couple of hundred years, and that everyone except 'grass and bamboo' would get out. And I have to laugh, of course, because bamboo is also a grass."

After dinner our little friend, one of whose charms was that she wondered about everything, asked me why I had joined the monastery.

I told her, "Some time in nineteen forty, when I was reading some of the sayings of Sri Ramakrishna in *The Bible of the World,* I had run into his suggestion, 'Try for three consecutive days with genuine earnestness and you are sure to succeed.' That sounded possible, so I tried. But at the end of three days I found that genuine earnestness was still a long way off. So I tried again. That was the end of my old life.

"That was several years after I had become convinced, by meeting Swami Ashokananda, that there must be something underlying the world which we see. But after those six days I became persuaded that there must also be something to do about it. That's when I wanted to join the monastery. That's when I took up spiritual practice seriously, and I wanted to know how to go about it. As I sometimes say, 'When a woman gets pregnant, she becomes interested in child care.' When you *need* the information, you go after it. In those days I listened very carefully to all of Swami's lectures.

"I was still living in my mother's house at ten nineteen Ashbury Street, but I had little to do with anyone there. I was out of the house before dawn, and off to meditate in Sutro Forest. I ate very little, and spoke less. If I slept, it was only briefly, flat on my back on the floor, lest in a longer sleep, I should forget where I was going. Up again, long before daylight, out into the dark, wondering - that's what it was, wondering! What's behind this darkness?

"In those days my hair was long. I no longer cared. In those days my foot was hurt. Crutches took me where I went, fifteen mountain miles at a stretch. Death would get me out of that.”

“So that’s why you have those crutches in your room,” she said.

“Yes. You see, I was an athlete. Once, in a Boy Scout track meet, I took first place in all events: high jump, broad jump, shot put, discus, the hundred yard sprint, the two twenty, and the four forty. That’s a quarter mile. Then I wasn’t tired enough so I did a half mile. So when my foot got hurt, and I thought I’d never walk again, it was a *total* blow to my way of life, and I thought, ‘Death will get me out of this.’ That was my consoling thought. But I became an expert on crutches, and took fifteen mile hikes with the rest of them. Then, after most of a year, I suspected that my foot was out of joint, and I kicked it back in, and it got well.

“I keep those crutches as a reminder, and I walk on them from time to time, partly to keep my hand in, and partly to remind myself of the exalted state of mind I had back then. Just walking on crutches picks my mind up.

“I was still on crutches when I went to Swami for guidance, hoping never to go home, and he sent me back to school. For a while I went to the University with long hair and crutches, and I carried my books in the big pocket in the back of my canvas jacket. But I was afraid of school for fear they’d teach me all sorts of worldly things.

“Swami said, ‘Knowledge isn’t worldly; *attachment* is worldly. The austerity of a student is to study. Go right after it!’ That was during the War, and Vedantins could not be deferred from the draft as conscientious objectors because of the Gita - because the story is placed in a battle field. But I’ll tell you a funny thing. When I first went to Swami for advice I gave him a handful of matches, and he wanted to know what that was all about. So I quoted one of their old books, ‘The disciple is to approach the teacher with fire wood in hand.’ I said, ‘This is modern fire wood.’ He laughed. I think he got a big charge out of that, and long later I heard him telling one of the other swamis about it.

“But I’ll tell you another interesting thing. Around nineteen forty, while I was still living in my mother’s house, something happened that would normally have made me very angry. But this was after my six days, and I was very cool headed. So I asked myself, ‘If I get angry, would it do that person any good?’ The answer was No. If I get angry would it do *me* any good? No. Would it do anyone else any good? No. Strangely enough, for the next five years there wasn't anything anyone could have done to make me angry.

"It's like what happened with my friend's dog. In the twenties and thirties our family lived on Thirty Seventh Avenue, and there was this Pit Bull that lived on Thirty Sixth. And every time we rode my brother's bike down that block we kept our feet on the handle bars so the dog couldn't get at us. He was a terror. Then he and his master, Jack Aratta, moved almost next door to us on Thirty Seventh, and Jacky and I became best friends. We went hiking together. We went to Scout Camp together. We did almost everything together, but the dog paid no attention to me. He paid no attention to anyone but Jacky. Whatever Jacky told him to do, he did, but there was no one else. It was like this for several years. Then, one day, I was sitting in the back seat in Jacky's car and Jacky was in the driver's seat, and the dog was in back with me. And Jacky and I were just sitting there talking, and the dog put his nose in the palm of my hand. It was a total decision. From then on there were two of us. Whatever I asked he would do. It was that way till the end. *It was a total and final decision.*

"That's when I saw that when the concentrated mind comes to a conclusion, it goes way inside. When Sri Ramakrishna threw the coins and the clay into the Ganges, the conclusion covered the rest of his life.

"And I'll tell you another one about me in nineteen forty. I was on crutches, and I had made my first appointment to talk with Swami in the morning. But I had determined that I would no longer handle money, so I couldn't cross the city on the streetcar. Well, that night we had a terrible rain storm, thunder and lightning, wind and rain. And, late at night, I was looking out the window through the storm, and I thought, 'Come Hell or high water, I'm walking across this city on crutches in the morning to see Swami'. Well, in the morning the weather had cleared, and I walked across the city on my crutches. And, strangely enough, for the next five years I wasn't wet by rain.

"I don't know how these things happen, but there's someone inside who notices them. And it makes it a little easier to understand Sri Ramakrishna's life.

"Oh, I forgot to tell you that Swami rescinded my determination not to handle money, but he warned me against it.

"I was diligent in those days, and my hand writing became very pretty. It's inspiring to me now just to find something that I wrote back then.

"Well, I had to graduate from the University and do war work for a while before Swami let me into the monastery. That was on Buddha's birthday in nineteen forty-four. And I came with the vow on my lips, 'for richer, for poorer, for better or for worse.' And it's been all of that."

"Do you know why he sent you back to school?" she asked.

"I thought then that it was so I could be deferred from the draft as a chemistry teacher. But later I've wondered if it wasn't to get this job done, putting science and Vedanta together. I know now that I couldn't have done it without finishing at the University. He may have wanted me trained to do this job. I had already done two and a half years at the University."

"It's only Advaita that will square with science." she said.

"Yes," I said, "I know." And I gave *her* a hug.

This earlier period of my life was almost unknown to Avila, and totally unknown to our friend. For the first year or two after leaving the monastery in nineteen sixty-seven, when I had no money, and no place to go, I lived as a guest in many homes, and hitch-hiked back and forth from San Francisco to Sacramento, ending up in San Francisco in a little basement room.

It was there that I was living when I first met Avila. For several years I had been teaching astronomy in several towns, in addition to Berkeley, and for several institutions, in addition to the University of California. And since I never spoke from notes, my mind was driven, every time, to organize afresh my understanding of the real world. That's how I got it straightened out, as much as I *did* get straightened out.

In the spring of 1976 Avila had enrolled in my class in San Francisco, and partly I suppose, because she had an 'outdoorsy' look about her, and partly because she seemed quite open and friendly, I asked her if she would like to climb Mount Tamalpais. 'Sure,' she said, and off we went next morning. She was a jogger, but willing to climb, and I must say, very good at it.

There are some people who, when you meet them, convey the impression that they have left something behind. They convey the impression that they are not quite all there. And there are others who, when you meet them, convey the impression that everything about them stands there before your eyes. Avila seemed to be all there.

In the summer of '76, after my classes were over, the two of us had gone with the San Francisco Sidewalk Astronomers on their summer tour to the Grand Canyon and Yosemite with their twenty-four inch telescope, thirteen feet long. Most of Avila's observational knowledge of the night sky dates from that trip, where, almost every night, she was busy operating one telescope or another and answering the innumerable questions asked of her by the visitors to the parks.

After the tour Avila had been busy, and I hadn't seen her much. It was therefore with considerable surprise that I received her phone call suggesting that I come and live in Berkeley. Shortly after she had moved into her new house, the other two women had moved out, and Avila needed help to pay the rent. And she'd rather have me to talk to than wait and take her chances on a stranger.

Avila's house, though not large, was roomy enough for three, and she offered me the big west room with a view out over the Bay. How could I refuse? As the house was close to the University, I had accepted her invitation, but only shortly before she and I had gone to the snows where we had found that girl and warmed her by the fire.


The Moon reached first quarter that night, and the girl was charmed to see it through the darkness of the wilderness.

"It seems to be floating in a sea of stars," she said.

The Sidewalk Astronomers at Grand Canyon

CHAPTER 7




## TO BERKELEY


In the morning I took the wheel. The girl sat in front, with me, and Avila took care of things in the back seat. We had broken camp fairly early and were away in plenty of time, through the rocks of the entrance road that had once been described by a European as 'only a rough place in the wilderness.' When we were out of the rocks and on smoother ground, Avila touched the girl's shoulder and asked,

"Tell us something that you take for granted and we don't."

"We take for granted" she replied, "that the notion of duality is a very different thing from the notion of plurality."

"In what way?" Avila asked.

"In that a relationship based on a duality remains provincial whereas a relationship based on a plurality becomes universal. Your romantic relationship is provincial. It gives rise to domestic problems, family problems. Your pluralities give rise to universal problems, to government, to the United Nations.

"It becomes important in physics when you consider the relationship between what J.D. calls your two All-American forces that operate on the inverse square law, electricity and gravity. The electrical field is provincial. It's based on a duality, what you call plus and minus, and spin-up and spin-down. The gravitational field is universal. It's based on a plurality. That's what Newton noticed, that gravity is universal, that the same gravity that falls an apple to the ground, falls the Moon toward the Earth and keeps it in orbit. And about this universal attraction of gravity Larison Cudmore says on page twenty-seven, of *The Center of Life,* 'How else do you explain that, if there were only two hydrogen atoms in all of space, they would "know" how to find each other?'"

"What a lovely remark!" said Avila, "The undivided *always* shows through."

"Yes," said the girl "and it was clever of J.D. to notice the Exterior Decorator's problem. If it weren't for the gravitational plurality the electrical duality of the hydrogen atom would collapse, and we wouldn't have atoms at all. And if it weren't for the spin duality we wouldn't have the atomic table, and gravity would collapse the planets."

"What do you mean, the planets would collapse?" Avila asked. "Why would *they* collapse?"

So the Little One explained that it was the spin duality that made the atoms so big, because the electrons can sit together only in pairs, spin up and spin down. And that's what prevents the collapse of the Earth, because you can't put two Fermi particles in the same energy state.

And the girl continued, "The electrical rest mass, as J.D. learned from Professor Lawrence in nineteen thirty-four is due to the squeezing down of a *single* charge against its *own* field. But it's also related to its separation from the *other* member of the *duality*, and the undivided shows *there too,* as the attraction between the plus and minus. But it's still a provincial problem, not related to the rest of the matter spread out in the gravitational field. The gravitational rest mass, on the other hand, is related to the dispersion of the whole Universe in the gravitational field. That's not provincial at all.

"Mach thought that inertia 'here' must be related to inertia 'there', that is, to inertia in the rest of the Universe, but he didn't know how. He didn't see that his inertia was the gravitational energy of the dispersion. He didn't see that energy and inertia are the same thing. That had to wait for Swami Vivekananda and Tesla, and for Mileva and Albert Einstein.

"That big brass pendulum, which you showed me in Golden Gate Park, before we went to Palomar, keeps track of all the rest of the Universe. The other half of its momentum can be anywhere in the rest of the Universe. I know, your books say that it holds the plane of its swing with respect to the 'fixed stars.' But you know, and the stars know, that there *are* no fixed stars."

"Beautiful!" said Avila, "You certainly see things in a brighter light."

"Maybe so," she said, and turned to me, "But you must remember that Einstein's geometry takes out field theory as well as wave theory."

"Wait!" I said.

That was in the car on the way to Berkeley. When we got to the town of Mount Shasta we were still ahead of schedule, so I toured them out past the fish hatchery to show them where Joe Bogart used to live. It was at his house that I was staying when I built the 'Kangaroo.' That's a pair of twelve inch binoculars, with a seven and a half foot focal length, that makes the bushes stand out from the mountains at several miles. The mirrors are more than two feet apart, and you have to make them look cross eyed if you want to look at something only two blocks way.

Well, I was just finishing the 'Kangaroo' at his house on the night when the astronauts first walked on the Moon. And I was like a little kid, running in to watch them on the TV, and running out to look at the Moon through the telescope. I knew I couldn't see them through the telescope, but I wanted to be looking at the Moon while they were there, while they were *walking* on the Moon.

And I noticed an interesting thing. When JPL asked them a question, you had to wait a bit before you got the answer. But when the astronauts asked the question, you got the answer right away. That's because when JPL asked the question, it took a little while for the question to get to the Moon, and another little while for the astronaut's answer to get here. And it sounded as though they were sleepy on the Moon. But you didn't even *hear the astronaut's question* till it got to JPL, and then you got the answer right away. I don't know how many people noticed this, but it looked as though they were wide awake at JPL, but not on the Moon.

The Bogarts had two dogs when I was there. The male was half coyote, and spent most of the year elsewhere. But male coyotes, as well as the females, take care of the pups. Well, he mated with the female and left, but he was back on the day the pups were born. Avila remarked that he must have had a very good clock.

Then I asked the girl if she would please repeat what she had said in the car about Einstein's geometry. She again pointed out that Thomas Young had done the double slit experiment in 1802, and had invented the wave theory to explain the observations. She said that Michael Faraday had discovered electromagnetic induction in 1831, and had invented field theory to explain *those* observations. Then she pointed out that, since in all these electromagnetic interactions, as well as in gravitational interactions, the space-time separations between the emission and absorption events are always zero, we don't need either wave theory or field theory to explain the observations. We need them only to make predictions.

I thought to myself that she takes the equations more seriously than anyone I ever met, and I wondered how Avila was managing.

The rest of the trip to Berkeley was uneventful except that the girl swam in the Sacramento River when we stopped for lunch. But, in Avila's bikini, she looked like a little girl trying on her mother's clothes. And safety pins were required to keep things in place.

Before we turned in for bed that night, Avila and I talked at some length about what had transpired earlier that day, but the girl did not join in those discussions. She was busy with my books, and asked me if I still had the tapes of Swami Ashokananda's lectures. I told her that I did, and showed her where they were.

All the next day I was busy making arrangements for classes, and, as we were no longer so concerned about keeping the girl hidden, Avila took her to the University library on the Berkeley campus. What a surprise! After looking up a few books, she apparently became disinterested and said, "So many books! Everything one needs to know can be put in a library smaller than yours. This is an *ocean* of books the mere sight of which acts like a candle snuffer on the enthusiasm of my mind. So many things I'm not going to read! You'd better take me home, lest the flights of the butterflies leave my mind. You have enough books there, and there are the tapes of Swami Ashokananda's lectures which I have not yet played."

"Tomorrow is Saturday," Avila exclaimed when I got home, "Let's get Jeffrey and climb Tam. We've been so busy with the girl's physics; we've scarcely asked her anything about astronomy."

Jeffrey was a Sidewalk Astronomer, one of the first three Sidewalk Astronomers, and we knew that he was back from Yellowstone National Park where he was working as a geologist. I had suggested, years ago, that he make arrangements to have a movie camera stationed on a blimp over San Francisco, and aimed at the city. It should be rigged to be triggered by a seismometer so that it would record when there's an earthquake. It could even be done from the top of the new Sutro Tower. Then, if the city goes down, we could sell the film and rebuild the town.

I called Jeffrey on the phone. He tried to back out, but I insisted. "Never in all your life," I began. "I'll come," He said. "Pick me up at my place on your way."

We went up the Indian Fire Trail to the big flat rock on the north side of the mountain, and talked. We took off most of our clothes, lay in the sun and talked. Jeffrey had a great deal of catching up to do. He had obviously been much charmed by the girl when he was introduced to her in the morning, and had expressed his surprise that 'any such thing existed at all.' So, despite the limited conversation that had taken place on the way up, he had a great deal of catching up to do.

For the first hour or so we tried to clean up Jeffrey's physics, as the girl had cleaned up ours on that first day at the lake. Avila and I tried to do it while the girl watched. And, as sometimes happens when we sunbathe on that rock, the Telephone Patrol flew by and dipped his wing. The girl was much amused when we told her that the pilot had seen us and that, from a plane, waving is done with the wing of the plane, not with the arm of the pilot.

Avila remembered what I had told her about the big telescopes on the girl's home planet, and about their weather, and that they selectively burned their forests 'before the rains.' And she remembered about the 'butterflies' and the fact that they had no seasons because their sun was almost on their equatorial plane. So Avila wanted to know what determined when they had their rains.

In answer to her question the girl said that their heavy rains were not annual, and had little to do with their inconspicuous seasons. Instead, she said, they were more tied in with what we would call the sunspot cycle. Their cycle, it would seem, is much more prominent than ours and peaks every seven or eight years with great spots, huge flairs and, from her description, beautiful auroral displays which can be seen from almost any point on the planet's surface. She said that their magnetic field was stronger than ours and almost aligned with their planet's spin; so the auroral displays were almost centered over their poles.

Their sun, it seems, spins once in about sixteen of their days (about ten of our days, she thought), so that when there was a very prominent spot group it would reappear at sixteen day intervals, giving rise to auroras, followed by rain. She said that their 'festival of lights,' during which they burn some of their forests, is in celebration of these auroras, and is held before them, 'when the animals come down out of the mountains.' She suspected that, since their atmosphere was clearer than ours it became more easily supersaturated with water vapor. *That*, she thought, was why their rains were triggered by the particle discharges from their sun.


Jeffrey asked her a bunch of questions all at once. He wanted to know about the age of their planet and about their continental drift. He wanted to know about the strength of their gravitational field, and how far back their earthquake records went. 'Do they have prediction and control?' And he wanted to know how much of their planet's surface was covered by water.

She took the last question first. "More than here," she said, "which makes the climate much more even, more like Hawaii."

"Like our late Paleozoic and Mesozoic times," Jeffrey said.

"Mostly ocean?" I asked, "Or shallow seas?"

"Both," she said, "But at the poles it's ocean, and there are many islands on the continental shelf."

She thought the surface gravity was much the same as here. "I feel the same," she said. And she thought that *our* geologists were probably ahead in earthquake records and prediction because, she thought, their earthquakes aren't so bad. She said their plates were bigger with much less continental drift.

"Anyhow," she said, "we don't build towns like yours, in such haste and indiscretion. Also, our mountains are not as high as yours. We have nothing like your Andes or the Himalayas. And we have no mountain as beautiful as Shasta."

"And we have no mountain as beautiful as Shasta."

"Volcanoes?" I asked.

"Rarely," she said. "We're not into 'continental drift,' but when those mountains do go off they dust up the atmosphere for years, and pretty up our sunsets. Normally our sunsets are clean, and we can see your 'green flash' almost any time except after the fires. Our word for after sundown means 'after the green'."

I asked about meteor craters, like the recent one in Arizona which Avila and I had visited, some time back, with the Sidewalk Astronomers.

"Those we don't have," she said. "Our system is without asteroids, and without outer planets. However, we think our early continental drift was started by some such thing."

She thought that if their sun ever had outer planets they might have been removed by the tidal interaction with some other star. Jeffrey wondered about that, in light of the near coincidence of their orbital plane and the plane of their planet's spin, and also the rarity of a close passage by another star.

She thought that their planet might be older than ours but that life may not have started so soon after its formation. And they knew of only one planet between them and their sun.

Remembering that, after visiting Meteor Crater in Northern Arizona, Avila and I had visited the Petrified Forest in the Painted Desert on our way to the Grand Canyon, I asked the girl if they had any petrified forests.

"Petrified forests and coal we have, like you," she said, "but the coal was used up long ago, so now it has the rarity of a museum specimen. Now we get our power from wind, from what you call windmills. Our continents are small and the winds are very steady, like San Francisco in the summertime."

"Fossils?" Jeffrey asked.

"Yes, fossils we have." she said, "That's how we know, what we *do* know, about our past. But no further back than yours, though our planet may be older. And we've had an oxidizing atmosphere for a very long time."

I asked Avila to remind me to show the girl the fossils in my room. Still packed in my closet, I had a beautiful fossil trilobite and a thin, flat piece of very dark rock, almost like slate, showing, on both sides, the almost white imprints of several ferns. Avila and Jeffrey, of course, had seen them at 1600 Baker Street before I moved to Berkeley.

On the way up the mountain I had told Jeffrey about the girl, and about our trip to the lake. I had told him about our discourses on physics and about the trip to Palomar that the girl had taken with Avila. When he heard about the 'speckles on the pictures' he said, "If they know galaxies like we know stars, maybe they know how galaxies are formed."

At the Summer Scientific Meeting of the Astronomical Society of the Pacific in Ensenada, Mexico years ago, where I met Dr. Margaret Burbidge, I had heard a lecture on galaxy formation by Dr. Gott. But I didn't like the fact that he had asked the computer to make a galaxy out of a mixture of stars and gas. Galaxies aren't made out of stars, I thought, they're made out of gas, or the entropy wouldn't go up. If you tried to make a galaxy out of stars, they would just miss each other and go on by. But gas clouds collide, and the entropy goes up.

So up here on the flat rock on the north side of Tamalpais, Jeffrey asked the girl if she had any idea how a disc galaxy like ours might have been formed.

She said, "J.D. knows some of that. But first let me quiz you. If there are breezes in the inter-galactic mix, which size clouds would you expect to form first, big ones or little ones?"

Jeff answered, "Big ones, of course, as J.D. says, because the gravitational pull on the particles goes up linearly with the diameter of the cloud."

"So," she said, "if we start with an enormous cloud, big enough to form a cluster of galaxies, where would the first galaxies form, and would they be big or small?"

"They would have to form near the center," Jeffrey said, "where the density is greater, and I suppose they'd be big like M86 and M87."

"Good for you," she said, "you've been in J.D.'s classes."

"Yes, he's worked on my head."

"Well," she asked, "what would happen to a smaller galaxy, or a smaller cloud, falling past one of those bigger ones?"

Jeffrey answered right away that it would get spun up. It would pick up angular momentum.

"Yes," she said, "and the stars that formed in that spinning cloud would transfer that angular momentum to what J.D. calls the 'hovering layer' by what he calls their great magnetic eggbeaters. Stars that form in a spinning cloud would gather up the angular momentum of the cloud. And because the stars shine in the ultraviolet and knock the electrons off the atoms nearby, their spinning magnetic fields would transfer that angular momentum to those charged atoms which are then blown away to the disk by the stellar winds of those stars."

"Yes, "I said. "I got that eggbeater idea from Hannes Alfven long ago. When the stars spin, their magnetic fields *must* spin with them, and they transfer their angular momentum away to the electrical particles near by. But if they didn't shine in the ultraviolet and knock the electrons off the atoms nearby, they couldn't transfer that angular momentum away. They would have eggbeaters, but no eggs."

Then something blossomed in my mind, and I continued.

"So *that's* how the angular momentum of even the central bulge stars gets transferred to the disc. It gets transferred to the electrical particles that get blown away to the hovering layer by the stellar winds. That's why the disc of NGC 4565 is flat whereas the central bulge is not. Almost all of the angular momentum of the early stars, in both the globulars and the central bulge, must have been transferred to the hovering layer, along with the dusts of the exploded stars, and it's the *original* angular momentum of the cloud that flattens the hovering layer into a disc. Bravo!"

"Yes," she said, "and the stellar winds of the central bulge tend to clear the central bulge of gas. It goes out to the hovering layer, and only in a disc like yours do stars have enough angular momentum to have planets orbiting around them. That's how I know that *our* star must have formed here in this disc."

Jeffrey and I were sorry to hear that planets could form only in the discs, because he and I had thought that there might be planets like Jupiter and Saturn even in the central bulge. We knew there couldn't be planets like the Earth and Mars in the globulars or in the central bulge because those stars lack the heavy elements out of which the Earth and Mars are made. But we thought that they might have planets like Jupiter and Saturn, made of lighter materials.

When we mentioned all that to Avila and the girl, and referred to Jupiter and Saturn as planets, the girl said, "Oh no, Jupiter and Saturn are stars. They spin too fast to be planets. They must have come down with the Sun, all three of them spinning fast. But only the Sun was big enough to shine in the ultra violet and knock the electrons off the atoms nearby. Only the Sun had electrical particles nearby to slow down its great magnetic eggbeater. Jupiter and Saturn are still spinning fast. Probably some ninety percent of the angular momentum of the Solar System is now in them. Most of the angular momentum of the early Sun was transferred away, back again to the hovering layer.

"And the Sun isn't hot because of nuclear fusion as they say in J.D.'s books. Hydrogen fusion is driven by gravitational collapse. It's gravity that makes it hot. Nuclear fusion is simply a delay mechanism on the gravitational collapse because all that energy has to leak out to the edge of the Sun to get away, and it must take about a million years to get from the center to the edge."

Well, I thought, if we can't have planets in the globulars and in the central bulge, then we can't have them in M86 and M87. We can have them only in the dusty discs of spinning galaxies. 'What a come-down!' I thought. That reduces the number of likely planetary systems in the Universe to a small fraction of what I had expected. I'm going to have to change a great deal of the things I say in class.

We ate our lunch. Jeffrey was sunburned. Avila and I were too brown to burn, and anyway we had taken our PABA (Para-amino-benzoic acid), and the girl seemed immune like Urania's kid, Uranio, who was half Mexican.

Putting on our shoes, and some more of our clothes, we went down the mountain on the other side to show that inquisitive girl the pretty green serpentine, mixed with brown, south of the west peak of the mountain.

Always, when one has something to show, there is the hope that the one to whom one shows it will love it. But rare indeed is the opportunity to show something to such a one as she. She loved everything, and we were like kids showing things to their mom.

And she said, "Showingthroughness is what we see as beauty. Showingthroughness is what we see as love. The Underlying Existence *always* shows through.

"The beauty of the eternal snows atop the craggy cliffs is the changelessness of that Underlying Existence showing through. The beauty in the freedom of motion of the eagle on the wing is the infinite showing through. And the beauty of Juliet's plea is the undivided showing through.

> *'Come, gentle night; come, loving, black-brow'd night,*
> *Give me my Romeo; and, when he shall die,*
> *Take him and cut him out in little stars,*
> *And he will make the face of heaven so fine*
> *That all the world will be in love with night,*
> *And pay no worship to the garish sun.'*

That's the undivided showing through."

We scrambled down the serpentine slide, smelled the fragrant leaves, the mint and the bay, drank from the stream, and we were back in Berkeley before the Sun had set.

Jeffrey agreed to stay overnight, and after dinner, I think mostly for his benefit, the Little One told us who we are. She saw us as an ancient race, come down through a long genetic past. And we found it both delightful and uplifting to see how we look to someone from outside, to someone from so far away.

"For hundreds of millions of years you have been bullied and pushed around," she said, "driven from the ocean to the rivers, from the rivers to the shallows, from the shallows to the swamps, and out on land. Always the species that were better adapted to the older environment stayed in the older environment. The faster fishes stayed in the sea. You are not descended from them. You are descended from a long line of misfits who were bullied and driven out. Always it was 'shape up or get out,' and you got out.

"You were driven ashore on stumpy fins in the Devonian swamps, and you were driven underground in the Paleocene grass, and you were driven from the grass into the trees, by other descendents of those stumpy fins. And every change entailed millions of years of discomfort while you painfully built in your new genetic adjustments, not so much by the survival of those who succeeded as by the early demise of those who failed.

"The dinosaurs, with scaly feet, drove some of you underground. Those who couldn't adjust are gone. There in their burrows, in the sunny grass, the rodents, furry mammals much like you, but better adapted to the grass than you, drove you to the trees. Those who couldn't adjust went down.

"There in the trees, through long and painful genetic readjustments, you learned to swing from branch to branch. Those who failed were eaten by cats. Then, after many more millions of years, just when your arms could reach from side to side, came the dwindling of the forests by drought, twelve million years of drought. Those who were better at swinging than you drove you to the ground, and you fled to the sea. You had four hands and no feet, and the grass was now no place for you. There were pack hunting dogs and great stalking cats. Those who didn't make it to the beach are gone.

"In the safety of the terrifying breakers you were cradled in the sea with hands instead of paddles, and hands instead of feet. And there were millions of cold, wet, salty years before you even had the tears to cry. You were small and you were timid when you came from the green-roofed jungle with eyes accustomed to the dark, and there were millions of years of blinding brightness on the sunlit waves and beaches before you had the frown of your bewilderment, the furrowed brow of the thinker, and you wondered what it's all about.

"The long pursuit has made you thoughtful. Every new adjustment entailed a genetic enlargement of the brain. It is the brain of a misfit, driven hither and yon to the refuge of new environments by those better adapted to the old. It's the brain of a shiftless outcast living always in the discomfort of genetic maladjustment. It's the product of hundreds of millions of years of distress, the product of the vicissitudes of countless misfortunes encountered along the seemingly endless reaches of the immense journey, as Loren Eiseley calls it. And your present form is not the end. The journey lies as far ahead as behind. No, not so far, for now, for the first time, you can look behind to see how you have come. And now, for the first time, you can guess ahead to see how you should go.

"In all that three hundred million years no creature descended from the Snout thought of himself as descended from the Snout, that lumbering Devonian fish with simple lungs and bubbles in his brain. In all that length of time no creature thought that any creature would ever think to figure it out, to unscramble and decipher the account. You are the first species that ever investigated its own genetic past. You are the only creatures who are not fish who ever knew that they are not fish but that their ancestors were. You are the first creatures who ever lived on land but who knew that their ancestors lived in the sea.

"And you are the first creatures who can look ahead to see where you are going. You are the first creatures who can understand that you got into this mess by an uncertainty and cannot possibly get out by transformation. Uncertainty is overcome by knowledge, not by transformation. You alone can understand that the journey has an end which cannot possibly be reached by journeying."

Here, at Jeffrey's suggestion, we had a break for graham crackers and milk, and then, to Jeffrey's great delight, the girl went on.

"Yours is the strength of the eternal underdog. You have been pushed and bullied and driven till you have mastered every environment on the face of the Earth and have the brain to comprehend the Universe beyond. Out of the endless vicissitudes of your misfortunes and your failures has come your strength, and your love for the underdog. Every unbiased observer among you roots for the underdog.

"When you walk in the woods the squirrels don't bring you their peanuts, but you carry peanuts for them. The gulls don't bring you their lunches, but you throw your lunches to them. And signs are required at every zoo to keep you from feeding the underdog. Out of the strength to save yourselves has come the strength to save others. You are Eiseley's star throwers, throwing the broken starfish back into the sea. Hundreds of millions of years of distress have gone into that strength, and the salt of those eyes.

"For hundreds of millions of years you have been bullied by the superior genetic technologies of better adapted species. You were hurt by the pincers of crabs, bled by the syringes of insects and killed by the syringes of snakes. You were scratched and torn by the talons and beaks of birds, crushed by the hoofs of mammals, tossed by their antlers and gored by their horns. Losing the sea to the fins of faster fishes, long ago, and to the flukes of faster mammals, only yesterday, you came ashore again only to be slashed by the fangs of cats, descended by another trail, another trial, from that same Devonian fish. Into every new habitat you came, you came lately. Everywhere you looked there was someone ahead of you. Everything you could do, they could do better.

"Every vicissitude of your misfortune had robbed you of some piece of genetic hardware which could have saved you in some niche, till, by the time you came, a second time, ashore, you had no fins, you had no flukes, you had no tusks, you had no claws, you had no hoofs, you had no fur. You were a ne'er-do-well's ne'er-do-well, protecting naked babies in the grass.

"Without pincers, without syringe, without talons, without beak and without wings you came ashore, with no trunk, no hoofs, no fangs and no fur. But something else you had. Behind your furrowed brow you had a better brain. Every single blow of your misfortune, which drove you to another niche and robbed you of some piece of genetic technology, had hammered on its anvil some improvement in your brain till you had now the gleam of knowledge in your eye. At the cost of losing every piece of hard-won hardware you have built the software behind your eyes. You have a brain to wonder and to understand.

"And you have breasts to feed the growing brain of your helpless offspring. And you have tools, and you have words to tell your offspring how to use them. And you have fire to protect both your breasts and your infant from the bullying of furry beasts with fangs and claws and chattering teeth. Only in your nakedness have you lost your fear of fire, driven by the cold and by your terror of the hardware of other species. Your every misfortune you have turned to your account. Through the unfortunate necessity of prolonged parental care has come the growth of that brain that uses fire. Only through the prolonging of your youth has come your wisdom which began in the swamp, long ago, around those bubbles in your brain. You are the descendents of that ne'er-do-well, air breathing fish, and the children of children who never grow up.

"Now, for the first time, you have a software technology before which all the genetic hardware has gone down. Now, with non-genetic hardware, you out-swim the fish, you out-run the cats, you out-fly the birds, and you look down from the Moon, and you smile. Just think what went into that smile!

"You have been pushed and bullied till you can be pushed and bullied no more. Every time you went down before the onslaught of some piece of genetic hardware you have come back with some unexpected improvement in the software behind your eyes, till now, with your software technology and the use of non genetic hardware, you, the eternal underdog, can bully any species that ever bullied you.

"But with your new-won strength has come the frown of your puzzlement, the salt of your tears, and your smile. Why should dog eat dog? Why should a species, once bullied, bully back against the species that bullied it? The furrowed brow has noticed and the salty eyes are wet. You are the underdog's underdog, and now that hand, once fin, once paw, lengthened for swinging in the trees, and flattened for swimming in the sea, now that hand, grown old, reaches out to touch, in consolation, those who in the past have bullied it. Was it not their bullying that made you what you are? You are the Star Thrower, throwing the broken starfish back into the sea. Save the condor! Save the whales! Save the leopard! Save the shark! Save that menace of the seas against whose fearful jaws you learned to clench your fist!

"You are the only creatures who ever knew that the rest of the creatures are just like you. You are the only creatures to have figured it out, that you got into this plight through an uncertainty, and can't possibly get out through a transformation. Knowledge is the key. 'There *is* no other path to go.'

"You are the first creatures to have figured out that the entire Universe is made out of hydrogen, but that the hydrogen is an apparition on the Underlying Existence seen in time and space. You are the first to see that your bodies, and every single terror that beset them in the past, arose, by transformation, from that colorless, odorless, tasteless inter-galactic hydrogen which couldn't possibly have arisen through any transformation, but only by the appearance of pairs of opposites on an underlying identity. And you are the first to see that only on that identity rests your concern for other creatures.

"That ancient, bullied hand still reaches out. That ancient furrowed brow has understood. And now the strength of knowledge lights those salty eyes. The end is not far, and, to one who sees beyond the transformations, the end is already within reach. The journey has been immense, and, in its immensity, it has yet to run, but the journey has an end which cannot possibly be reached by journeying."

Later that evening the girl asked me if she could play my guitar. "Of course." I said, and I moved to show her how to play it.

"I know how to play it." she said, "I have watched you sing and play, and I have read your Songs of Orpheus. When your brother gave you that guitar, you changed the strings, and tuned it for your own use, so you would never have to do anything complicated. I can see how you do it.

I wanted to *give* her the guitar, but she didn't understand my intention.

"He wants you to have it for your own." Avila explained.

The Little One chuckled and said, "I have never been able to understand what you two mean by 'owning' something. I have not been able to discern any measurable difference in an object before and after you 'own' it, except that sometimes you keep it in a different place."

She thanked me sweetly and said, "I'll put it in there."

I had two such guitars, classical guitars really, but not tuned in the usual way. They were tuned the way Orpheus tuned his Kithara three and a half thousand years ago. I used mine as he used his, to accompany singing. And I called them 'retarded guitars' because they were tuned so that any one could play them without much musical skill.

Orpheus had only four strings on his Kithara, four *open* strings. There was no fingerboard back then. We'll say he had a high c, a low C and an F and a G between. He could play c, C and G for his tonic, the c and the F for his sub-dominant, and the G alone for his dominant. That's all he could do. But my guitars have fingerboards and two extra strings. I have another high c, and another F, an octave down. My lowest string, low F, is tuned to the 'wolf note' of the instrument, and I sing in *that* key. (It was actually an A flat on my old organ.)

One of my guitars had been given to me by my youngest brother soon after I left the monastery. The other one I had bought from a hippie girl for only five dollars, but it too was a gem. On both instruments the high strings had been replaced by a set of low strings, (wrapped strings, and designated 4, 5 and 6). And both instruments were tuned to the 'wolf note' of the instrument, (the note that resonates most loudly when you sing into the hole).

In the key of my 'wolf note' the two highest strings, together at the center, were tuned to the dominant like the low string to the left, and the string between was tuned to the second note of the scale. (That would be D, in the scale of C.) Both the low strings on the right were tuned to the tonic, (the 'wolf note' and the octave above). Then, with my thumb holding two strings on the fifth fret, I could play six strings on my tonic. And with my little finger holding two strings on the seventh fret, I could play six strings on my dominant. It didn't take much skill. And with my index finger holding four strings on the fifth fret, which was a little more difficult, I could have four strings on the sub-dominant.

That girl had apparently seen all this by watching me, and she took one of the guitars to her room.

# CHAPTER 8


# TEARS


Next morning, after Jeffrey had gone home on the bus, and after I had come back from my morning walk and done my little worship, Avila begged me to sing. "One of those old Sanskrit things," she said.

So I got my guitar and sang three verses of Shankara's Dakshinamurti Stotram. And I translated it for the two of them.

> *Seeing the Universe existing within himself, like a city seen in a mirror, yet appearing as if produced outside through Maya or illusion, as in sleep: To him, incarnate as the teacher, to him be this salutation!*
>
> *He who, like a mighty magician or some great yogi, spreads out this Universe which exists indeterminate like the germ of a seed, and is later on diversified by the differences arising through the notion of space and time: To him, incarnate as the teacher, to him be this salutation!*
>
> *He whose existence is the substance behind such fictitious notions as I am the body or I am the doer, who, having realized the identity of the seer and the seen is no more reborn in this ocean of worldly existence: To him, incarnate as the teacher, to him be this salutation!*

The girl turned to Avila and asked, "Did you *hear* what he *said!* Shankara has Einstein's geometry built into his *perception.* 'Seeing the Universe existing *within himself,* like a city seen in a mirror, yet appearing *as if produced outside* through Maya or illusion, as in sleep.' But Maya is the Sanskrit name for the *mistake of seeing things in space and time.* You see it *is* possible to see this straight."

We sang a good deal that day, and in the afternoon the singing was interrupted when a butterfly got into the room, and Avila very carefully caught it and set it free. The girl stood motionless and watched, and when the butterfly was released, she retired to her room in tears.

We had never seen her cry, and I wondered what had set her off. Was it Avila's keen concern for the welfare of the butterfly, and the extreme care she had used in catching it? Or was it the freedom she sensed at seeing that beautiful butterfly released to the outer air? Or was it, I wondered, related to the green leafy butterflies of her own world, to something in her own past, about which I dared not ask?

Avila and I talked while the girl was in her room.

"I know how you picked up your Sanskrit," Avila said, "but how did you run into Orpheus?"

"Well," I said, "While studying Greek philosophy in the monastery, I had become persuaded that the Greeks had borrowed the five element theory from the Hindus, either through Pythagoras or, earlier, through Thales of Miletus. And, wondering why the Greeks were so predisposed to accept things Hindu, I ran into Orpheus"

Orpheus was apparently one of the sons of Phryxus, born and raised in India, in King Parikshit's court. He must have brought the Kithara from India to Greece, and, with it, the major scale. The older music of pre-Homeric Greece was the music of the pipes, the Auloi, with equidistant finger holes. They were played two at a time in the same mouth, and the scales produced were very similar to our minor modes. With Orpheus and the Kithara, came the major scale and its accompaniment with open chords, on the dominant, the tonic and the subdominant. Out of the juxtaposition of these two musical systems, one from India and the other one native in Greece, arose the history of European music with its interplay of major and minor scales so familiar to us all.

In Europe the music has been elaborated by the use of a great variety of chord changes, whereas in India, where the harmony has remained simple, the music has been elaborated by the use of a great variety of overtones which we don't use in Europe. Those were the overtones which that girl and I would sing in those Sanskrit songs, songs very similar I suspect, to the songs of Orpheus.

Later that afternoon, after some more singing, that clever girl asked me if she could play my organ.

"Of course," I said. "It makes a better accompaniment than the guitar, but you have to pump it pretty hard. I think the bellows leak a little."

J.D.'s organ in San Francisco, years later

I had an old reed organ which I had bought for about thirty dollars
from a second hand store on Filmore Street, long ago. It was only three feet
tall, three feet wide and one foot deep. And you had to pump the pedals while
you played. But by the time I had cleaned it up, replaced the bellows, and
tuned the reeds, it was a great deal more than satisfactory for accompanying
singing. Each key played two reeds, an octave apart, and it made a lovely
sound. And I'd opened the top and bottom so the sound could get out.

The girl didn't need any instructions; she could see how it worked. And
as soon as she found what notes she needed, we sang for a while with the
organ. Avila loved it, and I had some more translating to do. Then I sang the
first line of something I had heard only once, in the Berkeley temple, years
ago. I found out later that it's from one of the songs in M's Gospel.

*Thou hast lifted all my sorrow with the vision of Thy face, and
the magic of Thy beauty has bewitched my mind.*

Unfortunately that's the only line that I remembered.

The girl was interested in Sanskrit, and pointed out that the state of
knowledge implied by what was taken for granted in that language was most
unusual. In Sanskrit you find one word for mass dispersed in space with its
consequent gravitational field.
The word *Akasha* means space, and it also means mass, what Swami
Vivekananda called 'the first principle of materiality.' And, as the first of the
five elements, it means the gravitational energy inherent in that situation.
"Most extraordinary!" she said.
Then I added that the only reason I had gotten the five element theory
straight was because my dad was a zoologist and he had pointed out that
the first perception by the ear was not sound but one's orientation in the
gravitational field, which we read through the saccule. Only if the first
of the five elements was gravitational energy, and not sound, did the five
element theory, make sense. We read gravity with the ear, kinetic energy
(as temperature) with the skin, radiation with the eye and electricity and
magnetism with the tongue and the nose. The five elements are five forms of

energy, not things like air and water.

My dad had told me that when crabs molt, they change the sand grains in their ears, and if, at that time, you give them iron fillings instead of sand, they'll put them in their saccules with their pincers. Then, if you hold a magnet over them, they'll throw themselves violently on their backs. So the perception of gravity is in the ear.

I knew all these things when I was a boy, and I learned later, as a chemist, that temperature, which is perceived by the skin, is simply a measure of the mean kinetic energy of the molecules. It was easy, then, to put the rest together. Protons taste sour, and the nose reads magnetic matching with the molecules which we smell.

"In some ways," the girl said, "the people who invented that language were several thousand years ahead of the rest of you in physics. They even had the identity of mass and energy way back then, which you didn't get till nineteen hundred and five."

"Yes." I said, "And even *that* we got from *them*.

"Surely," she said. "And it's only from reading your Vedanta books that I could understand how you understood so easily all those things we talked about up there at your brother's lake. But you two should give up the discrimination between matter and energy. It's all energy. There's nothing else. It's the Underlying Existence showing through.

"Swami Ashokananda never seemed tired of pointing out that our actions are not governed only by *Karma* (the effect of past actions). Instead, he seems always to have kept it clearly in mind that the primary motive which governs our actions is the vision of the real. It's his *Asti, Bhati, Prîya,* the changeless, the infinite, the undivided, as seen in this world of time and space that moves whatever moves. And the world itself *is* that Underlying Existence showing through. There *is* nothing else.

"Space is not that which separates the many, but that which *seems* to separate the one, and in that space that oneness shines. Therefore falls whatever falls. That's *Prîya,* 'dearness.' As Jane Field wrote in your book, 'That which is beautiful is always loved.' And space is not that in which we see the small, but that in which the infinite *appears* as small, and in that space that vastness shines. That's *Bhati.* Therefore shines whatever shines.

Therefore bursts whatever bursts. That's your dispersional energy. And, finally, time is not that in which we see the changing, but that in which the changeless seems to change, and in that time that changeless shines. Therefore rests whatever rests. Therefore coasts whatever coasts. That's *Asti*, inertia."

"That's the physics in a nut shell," said Avila, "with its connection to the geometry. That says a great deal."

"Yes," she said, "but it doesn't tell you what to do. That, you know, *we* didn't know. You have no idea how *lucky* you are. You have all those lovely hymns that you sing, and all those other songs, and all that Sanskrit. And all those saints.

"But no one 'back home' knew that the Underlying Existence could be addressed as a person. And no one knew that someone could *get there*. You people have no *idea* how lucky you are, but even here almost no one knows what *you* know.

"Sri Ramakrishna used to tell the story of three travelers who came to a high wall and wondered what was on the other side. One man climbed, shouted for joy and jumped over. A second man climbed, screamed for joy and jumped over. The third man climbed, saw what was there, and came back down to tell others.

"You have people like Buddha and Sri Ramakrishna who came back down to tell others, 'bahujana hitaya, bahujana sukhiya,' for the good of the many, for the happiness and the welfare of the many. When you read the life of Buddha you see that he was like that all his life. Even the climbing of the wall was for others. He didn't leave the king's palace and take to the life of a wanderer for his own sake. He was trying to find a way out for *others* from this world of sorrow.

"But J.D. you are right, there has to be *understanding*, or even the one who climbs the wall may become a fanatic. Swami Ashokananda was very smart. Experience without understanding often leads to fanaticism."

"It took me a long time to understand all that," I said.

And I told our little companion that we have trouble here because there are those who feel that the ultimate reality is beyond the possibility of taking human form, and even beyond the possibility of personality. And against that view there is, of course, this question: If the Underlying Existence is infinite, who are we to say what is impossible? Who are we to tell it what it can and cannot do?

"Who are we to fence the infinite?" I said, but our friend didn't know that finite in Latin means 'fenced,' so I explained.

"Thank you!" she said. "Swami Vivekananda said that nothing could arise by evolution that was not already put in by involution. If the Underlying Existence weren't infinite, from where, pray tell, could all these marvelous beings arise? *And the magic of Thy beauty has bewitched my mind.* Whence all this beauty if the infinite didn't show through?"

We sang that line with the organ, and we sang some more, a good deal more, and that quiet girl, apparently much moved, went to her room.

CHAPTER 9


## THE APPROACHING END


Next morning, when I had finished my meditation and arose from my seat at my shrine, I found the Little One standing at the doorway of my room. She stood there quietly with her unblinking gaze fixed steadily upon me. She had never behaved in that way before, and at first I was at a loss to understand it. Suddenly I remembered that she was anything but forward and had never crossed the threshold of my room.

"Please come in," I said. "You may come in whenever you like. There's nothing private about me, or about my things. I thought you knew."

Smiling at me, she stepped quietly inside the door and asked if she might sit in my shrine.

"Of course," I said. "Have at it!" And I retired to the kitchen.

J.D.'s Shrine in San Francisco, years later

After Avila and I had had some breakfast, I was off to San Jose with the morning mail. I had a class to teach, and I mention the mail because of its importance to this narrative. I had written to Urania in the West Indies about this girl, and the letter that had come in that morning's mail was from her, insisting that I write this story. "Start it right away!" she had said.

At first, I didn't think to do it. I am not much given to writing, and I never wrote anything big. And, although I had written some of the physics down, up at the lake, it was not in the form of a narrative. But I was down there for two nights, in Los Gatos, and, on the second morning, I awoke at four, nearly three hours too early to be up, without disturbing the people in whose house I was a guest. And, lying there awake in bed, my mind was occupied with the problem of how to present this narrative. And, also, I was concerned about that girl.

My mind, in recent years, had shifted the direction of its approach from a pursuit of the perceptual grasp of reality more toward physics and an intellectual understanding. Is it possible, I wondered, that her mind has shifted the other way?

It was her response to those songs that had made me wonder.

When I got back to Berkeley that evening I found Avila in her room, alone. The girl had been gone since early morning, alone in the hills, without taking any food.

"It hurts a bit, you know," Avila said, "not being able to do anything for one so sweet."

My mind ran back over my concern for that girl. What is she after, alone in the hills? And I thought to myself, "We have offered a dime to a princess. What a travesty! Love is the gift for a princess, something that wealth cannot buy. What have we to offer her? Knowledge she has. That's not her problem. And the things of this world she doesn't want. 'What shall I do with that by which I do not become immortal?'"

And I was thinking about all the rain that had fallen from her world to ours. All the gifts have come from her to us. What is the gift for that girl? What can we offer her?

We had yet to see what rain would fall the other way.

I turned to Avila. "What's she after, alone in the hills? What has she been doing while I've been gone?"

"Mostly she's stayed in her room if she's not in the hills, or in your shrine. Sometimes she sings with your guitar, or listens to tapes."

"Does she sleep?"

"A little, and she gets up very early and goes out after flowers for the shrine. She doesn't talk and she doesn't eat."

"That's not bad," I said. "Those tapes are interesting. I've heard two great lecturers in my life, Swami Ashokananda and Richard Feynman. One knew things from the bottom, and one knew things from the top."

"The girl knows both," said Avila.

"For sure," I said. "Is she in good health?"

"Yes, she seems fine, but she won't eat anything, perhaps it's the time of her period, and she won't speak. It hurts, you know, not knowing what's happening."

Avila fixed dinner for three, but with the fear that the girl wouldn't touch it. She hadn't touched food all day.

We stood around and let the food get cold, and finally, just as we sat down, the girl came in, solemn but cheerful.

Avila stood up, and with a friendly gesture, waved the girl toward her chair. It was to no avail. Solemnly, tenderly, the girl shook her head. Then, without a word, she stepped behind Avila, and, gently placing her hands on Avila's shoulders, indicated that she should sit down and eat her dinner.

Avila's eyes were crying, as were the muscles of her chin. Somehow she managed to eat her food, and the girl stood behind her chair the whole time playing with Avila's hair. It was a moving scene; even my eyes were wet. And I had the feeling that things were coming to an end.

When dinner was over, the girl helped me with the dishes. Then we all went for a walk in the night. And no one said a word.

Next morning the girl came to breakfast and, without a word, indicated by the gesture of walking her fingers across the table that she would go for a hike.

Avila was delighted.

"Let's take her to the blue, blue lake."

The blue lake meant Bon Tempe, one of the reservoirs on the other side of Tamalpais. There were four lakes on the north side of that mountain. Phoenix and Lagunitas were smaller, and green. Alpine was gray, a long, gray lake with steep, wooded hills on either side, and with a beautiful, tall concrete dam at the west end. Between Lagunitas and Alpine, over a vast field where we used to play Capture the Flag when I was a boy, was now this beautiful blue, blue lake, Bon Tempe. The water was very clear, and that was the only lake of the four, out of which Avila and I dared drink.

To this beautiful lake we took our little friend.

Lake Lagunitas

The lake isn't something we drive to. We reach it from Mill Valley on foot, from the south side, over the top of the mountain, or over the side. At my suggestion, we held mostly to the fire road over the side, so we could walk abreast and talk.

Well, we hardly spoke at all. We were past Lagunitas and almost to the blue lake before the girl said a word. Her mind had been elsewhere. She had been cheerful the whole way, but silent, and I wondered what was going on inside. I remembered that line, from the Gita, "Truthfulness is the austerity of speech, silence, the austerity of the mind." And I wondered what was going on in the silence of her mind.

Back in the early forties I would go for days on end without speaking, sometimes without eating also, and often without sleep. I was trying desperately to undo the body idea. One of the girls at the Unitarian Church noticed that I did a great deal more thinking than talking. In those days I was desperate to see through this. I felt that there was an ocean of joy behind what I could see. I could almost feel it, and I *knew* it was there. And now I wondered about that girl. Where has *her* mind gone?

One night a week in the monastery we listened to classical music on the record player. Swami said that when you transplant a plant you take some of the soil with it. Apparently he felt that we Europeans needed some European soil around our roots. Anyway, I always loved those music nights.

And always in the monastery, when I listened to music, I would feel that I had renounced everything and had wandered off, and, like the Pandavas on their final pilgrimage, I will never look back.

Just before we got to the lake, the girl said quietly, *"Thou hast lifted all my sorrow with the vision of Thy face."*
What has she seen? I wondered.
"Don't you see," she said quietly, "When we see that we're not the doers, not the managers, when we see that Mother is the manager, then and then only do we feel that all our sorrow has been lifted."
"That," I said, "is why I love that song." But the Sweet One didn't go on.

Usually we feel that our actions have some bearing on the future happenings of this world, and we forget that Someone Else is managing. We forget that we are not responsible for what happens in the Great Elsewhere. When Swamiji saw the ruins of a Hindu temple, destroyed by the Muslims, he was thinking about what he would have done, had he been there. That's when he heard a voice asking, *"Do you protect me, or do I protect you?"*

We lay by the lake in the minty grass on the north shore and had our lunch. There was a great deal of squawking from a pod of gulls out on the lake, far from shore. There were a couple of dozen together on the water and, occasionally, one gull would fly in or one would fly out.
Lying there in the sun, I was thinking of the problem of the interpersonal relationships for children raised by more than two parents. Even on Earth, older daughters often raise their younger siblings. And I helped raise my youngest brother. He was eight years younger than I was, and I was his protector. So I was thinking about that girl.
Her relationship to us had changed. When we had found her, she was like a child, and we were like her parents. We had rescued her from the snow bank, we had warmed her, we had fed her, we had clothed her (though we were never sure she cared), and in other ways we had served her where we could, as if she were a child. But when it came to imparting knowledge, it was the other way around. We were now the children. And when Avila and I had discussed this, we had the funny feeling that that girl, as mother, had spent

one whole year in finding out where we stood, and where we were going, so that in two short weeks she could show us, from our own point of view, where, for us, the forward direction now lay.

It was when we had taken the girl to my brother's cabin at Cliff Lake that Avila had first noticed this change. It was she who first noticed that whatever knowledge we imparted to that girl was used by her as an aid in imparting her knowledge to us. And I had noticed in those long conversations, when we talked long together without clothes, that sometimes her nipples would become more prominent, as mine do when I'm cold. I had noticed it first at the small lake, above the cliff. The weather had been warm and yet, when she had been talking to Avila, her nipples had become quite noticeably enlarged, so that their prominence had reminded me of the Mother Goddess depicted by the Minoans, covered head to foot with clothes, as I recall, except the arms and the full, round breasts with prominently enlarged nipples.

Remembering how prominent the girl's nipples had been that day in the warm sun, I asked her about it when we started up the road over the side of the mountain, on our way back.

"Yes," she had said, "and if either of you had touched my breasts at *that* time, by now they'd be in milk. Children don't belong to the parents you know; the parents belong to the children. And at that time I saw you two as children, because I knew some physics that you *needed* to know. And you must remember that in our world milk may be available from more than one mother. Our hormones aren't triggered like yours. Ours are triggered differently.

"But now I see you again as I saw you when you rolled me in the warmth of that fire in the Sierras, because now I see that you have something that I don't. Your people know how to see through this mistake. Knowing where the physics goes doesn't get you there. *You* know how to get there. That's the name of the game, and your people know the rules."

Noticing that the girl was now willing to answer, I began again, but still as before, as a child, asking his parent.

"If all this is a mistake," I asked, "in whose mind does the mistake arise?"

"First," said the girl, "this is not a mistake that has taken place in time.

Time is part of the mistake, as you now know.

"Second," she said, "seen from the standpoint of one who sees the mistake, the question: *In whose mind has the mistake arisen?* does not arise.

"Third, from the standpoint of one who does *not* see the mistake, the question: *In whose mind has the mistake arisen?* does not arise.

"There is no standpoint from which that question can arise, because mind is part of the mistake."

Avila stopped in her tracks and looked back, and the girl continued.

"However," she said, as we slowly resumed walking, "the Vedantins say that if, in that oneness, twoness is seen, then the perceiver must be one of those two. There is no other way. That twoness, they say, is seen in *'consciousness space.'* That's the space of what the Vedantins call *Savikalpa Samadhi. Shiva* represents the changeless, and *Shakti* represents the undivided and the infinite."

"But why the terrible aspect?" I asked.

"Because Shiva-Shakti is the *perceived, not* the *perceiver.* Infinitude in the *perceiver* is seen as freedom. Infinitude in the *perceived* is seen as terror.

"In the *'mind space,'* they say, the perceiver sees himself as one among many. That's the space of our interpersonal relationships. There the changeless is seen as peace, and the undivided and the infinite, as love and freedom.

"Here in what they called the *'great space,'* what J.D. calls the Great Elsewhere, that is, the space which seems to separate the stars; we see twoness in the electrical field, and manyness in the gravitational field *without* being one of the two *or* one of the many. Here the changeless is seen as inertia, and the undivided and the infinite, as gravity and electricity.

"When you dream, there's a space in your dream, and many things happen there. But when you wake up, you see that the separation between you and the dream was zero. It's the same in the waking state. And there's *always* a dreamer. That's why the Hindus believe in rebirth; *the dream cannot destroy the dreamer.*

"Quantum mechanics, the double slit experiment, and Feynman's 'sum over histories' are the observational evidence that the geometry of what is known in the trade as the real world is four dimensional, and that space and time come into that geometry as a pair of opposites, so that the space-time separations between the emission and absorption events for what are known in the trade as 'photons' and 'gravitons' are zero. That allows us to see, by mistake, a Universe as if spread out before us, yet with zero separation between us and what we see, and with zero separation between us and what affects us by gravity. It's like a dream.

"Gravity, electricity and inertia are the observational evidence that we are seeing, in time and space, an *Underlying Existence* which is *not* in time and space, and is therefore neither changing, finite nor divided. The changeless shows through in the misperception as inertia; the infinite shows through as the electrical energy of the minuscule particles; and the undivided shows through as gravity and the attraction between opposites. That allows us to see a Universe of hydrogen, dispersed through space and falling together by gravity to galaxies and stars, planets and people. It's like a dream, and the notion that we're made out of meat, as Dorothy noticed in the *Wizard of Oz*, is part of the dream.

"The life of the Old Man in J.D.'s shrine is the observational evidence that the Underlying Existence may be addressed as Mother, and that it's possible to reach Her through many different paths. That allows us to see that it's possible to awake from the dream. There's only the Underlying Existence. Everything else is part of the dream. And, for me, this last part, that it's possible to reach the Underlying Existence, is *new*. Forgive me! But *that part* is new. This Universe has an escape route which no one back home had known.

"Our people know where the physics goes, but not how to get there. And even here only a few know how to go about it. Most people are busy just playing with their toys. They have no idea what it's all about.

"We live in a world with an escape route, like one of those runaway truck ramps beside the highway, filled with gravel, but this one's not filled with gravel."

Her point of view is changing, I thought. Does she see that Underlying Existence where I see sticks and stones? Does she sense what I used to feel, that there's an ocean of joy behind what we see?

Back then, when I lived in my mother's house, I used to look out the back window at night and think, 'If all those houses burned down there wouldn't be as much sorrow as there is there now.' But also, looking out that back window at night, I used to feel that there was an ocean of peace and joy beyond all that I saw. And if it weren't for that feeling back then, I never would have joined the monastery. And, back then, I wrote in a letter to Swami, 'If there's any difference between calmness, serenity, freedom and bliss I haven't found it out.'

Once, on the way back from the lake, when we dropped below the road for water, we poured water over each other's topsides beside a pool. Here the water was very cold, and after I had poured for them, the Little One poured for me. I didn't know then that the three of us would never wash that way again, but now I think the Little One knew.

On the way down the southeast side of the mountain, Avila stopped to draw our attention to the blossom of a Sticky Monkey flower beside the road. She pointedly called our attention to the particular shade of its color. Sort of apricot, I suggested, or cantaloupe, but it was obviously neither. And I wondered what it was, in that shade of orange yellow, which had so captured and held my friend's attention.

I mention the incident because it was her drawing our attention to that flower that drew my attention to something else, namely, how much we take for granted in almost everything we think or do. The problem had become important in our conversations with the girl at the lake in the mountains, because on so many occasions we had failed to understand, or even to follow her, simply because we had taken something for granted.

There are things we take for granted in our genes, things we take for granted in our language, things we take for granted because of the way we were brought up. And there are things we take for granted because of our sheer familiarity with them in our surroundings.

That Sticky Monkey flower is common on that mountain, growing everywhere, and we had seen it all along the road. How rarely, I thought, do we ever look back at something, with attention, when once we have taken it for granted.

I was reminded of an incident which had taken place, not far from where we were just then walking. It had happened on that mountain two and a half years ago, barely a month before we had found that girl in the snow. It was already winter, after Christmas, and the Bay Area had been forty-five days without rain. It was an overcast day, and it had already started to drizzle by the time Avila and I had started up the mountain. But because it was rainy, we decided against trying to make it to the top, and instead we took the Echo Rock Cutoff, and had our lunch at the Rock. That trail was new to Avila.

Now the other end of that trail comes out at the junction of the Big Bend trail and the fire road, and through that junction Avila and I had several times walked, both up and down. But approaching that junction, from an unfamiliar trail, she had failed to recognize it. Suspecting that it might be so, I had pointedly asked her if she knew where she was. "No," she had said, "I've never been here."

I took her down the trail to get back home by a different way, and from time to time I asked her if she knew yet where she was. She soon came to know, from my repeated questioning, that she must be on familiar ground, but, try as she might, she couldn't place it.

For the next hour or so she was like Alice in Wonderland, looking at everything with wondering eyes. She could recognize some trees, and the pools where we had drunk from the stream or splashed each other with its water. She could recognize the log, which formed a bridge across the creek and the big, green, mossy rock where, on a former occasion, the two of us had stopped to talk.

At the bottom of the Big Bend trail, to avoid breaking her spell by taking her to the Old Railroad Grade, which I knew she would recognize, I took her up a side trail, also familiar to her, to the small road leading up to the ridge. That's the ridge along which we had just come, only an hour and a half before. Along that road we soon came out of the trees and under the open sky, with the top of the mountain hidden in cloud, but with wooded ridges all around in the bright rainy haze.

It was all a wonderland. She looked at all the plants, the trees and the ridges in amazement. Everything was 'new' and yet strangely familiar. And, with her arms raised, she wheeled round and round in her amazement.

When we reached the ridge she knew that she had been there before, but she had no idea where she was. And although she knew that she had been there, she had no idea how long ago she had come by that way. For a moment she thought she was on the Throckmorton, but she couldn't square that with what she saw.

Where the road along that ridge meets the road that cuts through it, she had once before suffered such a confusion, but only briefly, when I had brought her in by an unfamiliar road. When we got to that point, this time, she saw the whole thing, and knew then that she had been by there, at that very spot, only an hour and a half before.

We had laughed at the fun then, and talked at some length about how much we take for granted, and of how the wonder of that last hour would not have been possible for her if she had 'known' where she was. And yet the wonder would have been there all the same. The wonder is with us all the time, I had thought, "but only he who sees takes off his shoes."

I reminded Avila of her wonderland and of the fun we had had, and we laughed again. And she told the story to the girl. Then we showed the girl how we sip nectar from the Sticky Monkey flowers, and then I told them both about the old railroad with the hinged engine that used to pull the 'gravity cars' up from Mill Valley to the Tavern near the top of East Peak. From there, I told them, the cars were allowed to coast down to Muir Woods, without the engine. It must have been a very scary ride because, when our family was hiking on the tracks, we were warned of the oncoming cars by the screams of the passengers.

Through all this, we got down to where the car was parked beside the Old Railroad Grade in the north end of Mill Valley. Then it was over the bridges to Berkeley.

That evening, in Berkeley, we sang with the organ, and at the girl's request we sang over and over again,
*Thou hast lifted all my sorrow with the vision of Thy face, and the magic of Thy beauty has bewitched my mind.*

"From hydrogen to bird wings and butterflies," the girl said rather quietly, "is by *Parinama,* but there's someone underneath with Her finger in the pie. And we don't go back by *Parinama* (transformation). We go direct to Her. All this is Mother, and if we don't want anything else, we get Her. The Old Man stayed here through all that suffering only so as not to upset the devotees. It's the love of a mother for her children. After he passed away he told Holy Mother (Sarada Devi) that he had just gone to the other room."

Before we turned in for the night, that pretty little girl asked me if I had ever noticed that Swamiji had passed away on America's Independence Day, that Holy Mother had passed away on France's Independence Day (the day the French people took the government into their own hands), and that the Old Man had passed away on India's Independence Day long before it happened.

"Yes," I said, "I noticed long ago. But Swamiji didn't die. He wasn't ill. He just left."

"Yes," she said, "I know."

"And," I said, "two days before he left he said to Swami Premananda 'Baburam, I want to go to every nook and corner of this world and tell them they are Brahman but the body gets in the way.'"

# CHAPTER 10

# FLOWERS


Next morning when I came in from my morning walk with flowers and fragrant leaves in the right hand pocket of my parka, I found the Little One standing by the alcove of my shrine. She knew, of course, that I kept that pocket just for leaves and flowers for the shrine. And she knew that at that shrine I did my morning worship.

And I had told her, some time back, that during World War II, when we were doing war work on the flood plain of the Sacramento River, in a cow pasture near Davis, that I had lived in a tepee that my mother had made for me out of unbleached muslin. In the middle of that tepee I had a coconut shell full of water, and each morning I offered my flowers at that shell. That was my shrine.

I had pitched that tepee with poles, which I had split with my hatchet, from a twelve foot plank which had been left there by the river in flood. And when the boss of the project told me that I'd have to go to San Francisco for three days, I was all pushed out of shape because there wouldn't be anyone there to offer flowers. It was all very serious for me then. But I had no choice; I had to go. Well, when I got back, I found that the wild Morning Glory, twining around the coconut shell, had three white flowers in bloom over that shrine. I had been gone for three days, and there were three white flowers in bloom over that coconut shell. Well, someone inside notices these things, and I didn't need to tell that sweet girl how I took it.

So this morning when I came in, I found her standing by the alcove of my shrine, and the Little One asked me very sweetly if I would please teach her that worship.

What is this? I thought. I have never taught the worship to any one but Ruth Ballard. And that was long ago at 1600 Baker Street. There, I lived in the basement and Ruth lived upstairs. But she did a lovely worship. She composed songs, and sang them in the worship with her guitar.

> *Born as you are on a rock near a star,*
> *He is here. He is near.*
> *All around, look around!*

> Or:　*Now has come the full turning of the wheel.*
> *Now has come the time to give up the unreal,*
> *For the man who dies, before he dies,*
> *Does not die when he dies, - when he dies.*

I don't remember all her songs, but that was a very beautiful worship which sometimes I heard when I was upstairs.

Here on Earth we do worship to turn our genetic programming around. The Little One's programming is different, I thought. Why should *she* do worship? So I asked her, "Why would you indulge in such an illogical ritual?"

"Why do you?" she asked.

"I'm not sure," I replied. "It used to be very important to me. In nineteen forty, and even later during the first years of my monastic life, it was very important to me to be able to do something which had nothing to do with any one, or anything, in this world. I felt, from my own experience, that the reality was completely beyond all this that we see. And I was looking for a way to reach it. Actions rise and fall within this visible world and cannot reach beyond it.

"In my mother's house I offered leaves and flowers before that little brass image of Sri Krishna that's there in the shrine. I used that image as a symbol of that reality which I felt was completely out of reach. I had never seen anyone do worship. I had never seen anyone offer leaves and flowers. But when I was a boy, I offered leaves and flowers to my mother, and to the girl whom I loved. When I came to understand that the reality could not be grasped through interpersonal relationships, I simply redirected that old habit. But why should *you* feel like that?" I asked.

"Why should I *not*?" she replied.

"Because you already know what the score is. I *didn't*," I said. "You already know where your physics goes. You already know that whatever we see here is that Underlying Existence."

"To know in your intellect, is one thing," she said solemnly. "To see in your heart, is something else. Our whole race knows where the physics goes; *your race knows how to get there.* And that is how you go about it, through worship and meditation. Now I see what I had never seen before. And now I want to go *that* way.

"Please," she said, "It's time to try *that* way."

"But there are so many ways," I said. "Why worship?"

"Why did the Old Man in your shrine go that way? Why did *he*," she asked, "who was a monist among monists?"

"I don't know," I replied, "But Swamiji didn't go that way, nor do I so much now, although I did at the beginning. 'It don't make no nevermind' to me now whether I do the worship or I don't, because my whole life is dedicated to that. Probably Sri Ramakrishna went that way because that's the way of the majority. But he left Swamiji to straighten us out. 'Exit praying and laying flowers in the temples!' Swamiji said. And he taught out-and-out monism, Advaita Vedanta, because he saw that everything the Old Man did and taught was based on that, on monism, on Advaita.

"Swamiji said that in India they go down on their knees before the man who studies the Vedas but not before the man who studies physics. He said that it is utter materialism and superstition, and not Vedanta at all. He said, 'Science and truth is all the religion that exists.'

"But," I said, "There are as many paths as there are people. So if you really want to learn the worship, you may offer *these* leaves and flowers. I have picked them very carefully. And I'll show you how we do the worship."

So I told her how we did the worship in the monastery, in San Francisco.

I said that early in the morning, before dawn, we rose and showered.

Then, putting on our shrine clothes, we meditated in the shrine room for an hour or so, sitting cross-legged on the floor. By then it might be light, depending on the time of year, and it was my duty, then, to clean the shrines.

Ceremonial purity must be observed. One does not touch one's person or one's clothes with one's hands through the whole procedure. And there's a hierarchy in the shrines. Worship is done first at Divine Mother's shrine, then at Sri Ramakrishna's shrine, then Holy Mother's shrine and Swamiji's shrine. Whatever one does, one does in that order, whether it be removing the flowers from yesterday's worship, or removing the incense trays to clean them, or cleaning the pictures and the shrines, or whatever. It's always done in that order. And nothing is picked up with the left hand. If there are two candles, one picks up the first with the right hand and hands it to the left hand before picking up the second.

It might seem, from outside, that these rules are silly, but they strengthen one's awareness of things unseen, and we found that they contributed to our sense of the living presence in the shrine. The problem is to persuade the mind that there *is* an unseen reality behind the seen. And one uses the shrine, and all the activities associated therewith, as a handle for the mind, as a help in getting hold of it.

That's the effort in worship, to get the mind to hang on through thick and thin. It's easy when the mind is 'up', but when the mind is 'down,' it hangs on by its habits, and ceremonial observance is one of these.

When the shrines had all been cleaned, my next job was to gather flowers for the morning worship. It's a *lovely* thing, gathering flowers for worship. With a special basket, kept just for that, one walks around in the monastery garden gathering only the choicest blossoms, knowing, in great joy, that in the worship itself each flower will be offered as a symbol of some virtue, some strength or some beauty in one's life. It's your *life* that you offer. It's *yourself* that you offer in the worship.

The flowering plants had been picked out by Swami himself, so that we could have flowers all year round. If at any time the flower supply fell short, I would report it to him, and he would get more plants that bloomed at that time of year. And I studied a good deal of chemistry to know how best to feed them. We dug them, we fed them, we watered them, we sprinkled them, we talked to them, and somehow all the activities of one's day wound up in those flowers at the shrine.

Five ingredients are offered in that worship, as symbols of the five great forms of energy of which all this is made. They are offered back, in reverence, to That from which they come. You offer the snake to the rope, as it were.

The whole thing is done with understanding. Before you start the worship, you meditate on that absolute oneness. Then you beg leave to do your little worship, knowing all the while how illogical it is. But it's done with devotion all the same. Finally, at the end, you offer back the fruits of whatever you have done, with an apology for being stupid.

I told the Little One all this, and showed her what I do. I told her that I still offer perfume, flowers, incense, light and food, as I did in the monastery, but that I now use fragrant leaves instead of perfume, as I have ever since I left the monastery.

I gave her a new cereal box for a flower basket, for tomorrow (I now use the pocket of my parka), and showed her the new handkerchief that I use for cleaning the pictures and the shrines. (I have a beautiful scarf over the dresser top, and plate glass under the pictures.)

I showed her the matches and the incense sticks, and I told her that nowadays I meditate sitting in that chair, my mother's old chair.

I even gave her my beads and showed her how to 'tell' them. (When you get to the 'guru's bead,' at the tassel, you go back around. You never cross the 'guru's bead'.) And I told her that nowadays I repeat *Om Kalyani Rupam,* and that I usually do it with the guitar instead of the rosary.

And I was thinking: The Vedantins won't teach the worship to one who has not practiced chastity for at least several years. She has practiced chastity all her life, and, perhaps, for generations before that.

It was when she had asked me to teach her that worship that it crossed my mind how extraordinarily well qualified she was as an aspirant for the highest knowledge. She might possibly succeed where some of us had failed. She was intelligent, clever and strong. She was fearless, yet sensitive, and completely unattached to the things of this world.

I thought of the four qualifications which the Vedantins regard as essential if one is to succeed in spiritual life: The discrimination between the timeless and the transient, the renunciation of the fruits of one's actions (not being caught in the consequences of transformational causation), the six treasures (calmness, mental restraint, one-pointedness of mind, endurance of heat and cold and the like, putting up with the faults of others, and the enthusiastic faith that the job can be done), and, finally, the yearning for liberation, that is, the yearning to realize the Underlying Existence in one's heart. All these she has, I thought. She is lacking in nothing. So I turned my worship over to her, to our strange little guest.

It was with great joy that I turned my morning worship over to that girl. Who gets such a privilege? And it was then that I noticed that some rain was falling the other way, from our world to hers.

The rain must have started with the singing of those songs. That was when I first had noticed her emotional response. Someone inside must have listened to those songs.

She had said that to know something in your intellect is one thing, but that to know it in your heart is something else. I figured that she must have seen it in her intellect some time back, but that those songs must have reached someone inside.

She had said that her whole *race* knows where the physics goes, *our* race knows how to get there, and that that is how we go about it, through worship and the like. And she had said that that is how the Old Man in J.D.'s shrine went about it. What could *I* say?

When first we found her in the snow, she was like a girl who had been rescued by her parents. Later on, the situation had changed, but toward the end it was like it was at the beginning. Through the singing and the worship she was, once again, a child before her parents. And I thought, 'Children are much faster than grown ups at learning things, and she is, once again, a child, and so good at everything.' And I wondered what might happen if she went the way of the Old Man.

Before Totapuri had met Sri Ramakrishna, he had spent some forty years in diligent spiritual practice. And when he saw Sri Ramakrishna seated on the bathing ghat at Dakshineswar, he thought, *'There's a fit person to practice this Advaita,'* so he asked him if he wanted to do it. Sri Ramakrishna went into the temple to ask the Divine Mother, and returned with Her permission. Then Totapuri, whom Sri Ramakrishna referred to as Nangta, 'the naked one,' asked him to come early in the morning to be initiated into the practice of Advaita. He taught him how to meditate on the One beyond all changes. But the disciple complained that when he meditated, he had the vision of the Divine Mother and couldn't go beyond. Then Totapuri took a piece of broken glass and pricked the disciple between the eyebrows and told him to meditate there. And he succeeded.

That's when he lost all consciousness of the outside world, all awareness of space and time, all awareness of what we would see as his body.

'It's the fitness of the disciple that counts,' I thought. 'Are you ready to give up everything else?'

Then I thought of that girl. There is nothing in this world that she wants, and in those songs and in that worship she must have glimpsed what lies beyond. She already knew where the physics goes, but didn't know how to get there. Now I saw, the rain is falling the other way.

Yama said to Nachiketa, 'Talkers we have aplenty; listeners are scarce.' Something in that girl must have listened to those songs, and now the rain is falling from our world to hers.

*May the likes of her attain the fulfillment of their desires!*

The first day after she learned the worship, she stayed in the shrine till lunch time but she didn't eat lunch or dinner. And she spent the afternoon alone in the hills.
Avila was upset, and feared an impending disaster. Like any mother, she was upset when her Little One wouldn't eat.

We sang a little that evening, and that lovely girl, with tears in her eyes, retired to her room without a word.

That left me wondering. Tulsidas (the author of the *Ramayana*) said, "If, by clinging to truth, by humility (the ability to see greatness in others), and by looking on all women as one's mother, God does not become known, Tulsi is a liar."

I thought of that girl. She's completely beyond the reach of our genetic programming, and anyhow she now sees both of us as her parents. And I have never known anyone to cling so tightly to truth, nor anyone so anxious to see the greatness in others, and now there's been all that singing. Then I remembered Mirabai, who said, "If a prostitute teaching a parrot to sing the name of God becomes illumined, so can Mirabai."

"Lordy, Lordy," I thought, "That girl's 'above and beyond.'"

The second day, after offering the flowers, she stayed in the shrine till nearly dinner time, and she ate very little. Eating didn't seem to matter any more. I had been away part of that day, and, again, we spent the evening singing. She loved those old hymns. She managed the hymn book, and we sang them over and over.

She reminded me of my mother. When my mother was a little girl in China, she used to play the organ for church. But she was too small to reach the pedals. Someone else had to pump the organ for her. And on Sundays, on her piano, she played the hymn book all the way through. She was by far the best hymn player I ever met. And she could play them in any key.

My mother, long ago (in Peking, where she was born)
She was the daughter of Dr. H. H. Lowry, Advisor to the Parliament
of Religions in Chicago in 1893.

This girl too was almost too small to reach the pedals on *my* little organ, so I played. She opened the hymn book and I played the root notes. I'm no better on the organ than I am on the guitar. But all that singing took my mind back to the time when I was a kid in the Sunday school choir, to all that caroling we did in the streets at Christmas time, and to all that ecstatic singing that Swami used to do in the worships.

Swami wouldn't let anyone record his singing at the worships, but I drilled a hole through the auditorium floor and sneaked a microphone through there. Nick Toth sat there in the second row with the microphone on his lap through the whole worship and, with his recording gear in the furnace room; we recorded most of Swami's songs that evening. We were in fear for our lives, but I thank God that we did it. I just wish we had done it earlier. I have never heard anyone sing like that. When he sang, he meant it, and we heard it that way. That girl must have heard his opening and closing chants on the lecture tapes, and she may even have heard the tape that we recorded that night. It's the only tape of Swami's singing at a worship, and, for fear of trouble, Nick Toth wouldn't give me the tape till Swami had died.

Well, the three of us sang till very late that night. We sang 'Thou art my All in All' to Schubert's Ave Maria, and we sang 'Nearer my God to Thee' in threes and fours, and we sang a lot of other hymns and things, and late at night our guest retired to her room without a spoken word. And I suspect that what she felt was beyond the reach of verbal communication.


After her worship on the third day she stayed in the shrine till dark, so I didn't go in. But before lying down on the couch to get some sleep, (before driving to Sacramento in the morning), I stole a peek through the door to see that she's all right. She was there, without a stitch on, seated in meditation on the floor. And I thought, 'She's throwing *everything* off like an insect fighting its way out of entanglement in a spider's web.' And I wondered whether she would throw off the body too. I worried. I wondered.


The next two nights I was away in Sacramento, but I phoned to Avila. She had taken her meals alone, and said that the girl might *still* be in the shrine.

On my way to Sacramento I had been thinking about those yogis in ancient India, how they had invented all those fancy exercises and all that careful eating in the hope of getting the job done in *this* life. They wanted *out*. And I was thinking about the Little One. She had finished what she had to do for us, and it was with a touch of sadness that I realized she might not stay.

I was not, therefore, totally surprised to find, when I got back from Sacramento, that the girl *really was* still in the shrine. So we knocked at the door and quietly went in. We found her there, lying on the floor, without her clothes, and the floor was stained with tears. Her breathing had stopped, her body had cooled off, and she was gone. *Never, was I more sorry to be right.*

She had thrown off her body, with her clothes, and the butterfly had flown. "But, where," I wondered, "has she gone?"

Avila, through her tears, said, "She has taken the 'runaway truck ramp' and gone home."

"It may be so," I said. She had asked me once whether, if she passed away and got reborn, she would be reborn here or there. I had said I thought it would be there. Who am I to know?

Avila was weeping, but my eyes were dry. It was the same as when my father had died in my arms long ago. It was extremely beautiful. It was much more beautiful than a death scene at the opera. And I didn't weep then, at death; I wept later, remembering all the things I should have done for him, and now there's no redress. This time it was the same. And I wasn't even present when my mother died. This time it was the same. May God forgive us!

She had said that the Hindus believe in rebirth because *the dream cannot destroy the dreamer.* It may be so. And I had told her that I thought that *rebirth is always in your own genetic line.* If she is reborn, it may be there. Maybe she went home. Maybe she took something with her. She used to worry that the people 'back home' didn't know the *escape route.* They didn't know about the *runaway truck ramp.*

Flowers

She had said that her whole race knows where the physics goes but that our race knows how to get there. Now perhaps she also knows, and has taken that knowledge back home.

Now, at last, I saw that the rain, the *deluge*, had fallen the other way, from our world to hers. And she was gone.

We found a note which that thoughtful girl had left on the floor of the shrine before her departure.

It's impossible adequately to thank you for saving the life of this body in the snowdrift up in the mountains, and for exposing me to this information, this *sacred* information. And I can never thank you enough for exposing me to all that singing, and to the worship. We don't have those hymns back home, you know. It's impossible to thank you, so I leave it to Mother. I found the runaway truck ramp.

What we see here is the Underlying Existence showing through. There's nothing else, only Mother. And only the hydrogen arises by the mistake. Then the Undivided falls it together, the Infinite makes it shine, and the Changeless makes it recycle. Nothing else is going on here except the runaway truck ramp. There *is* a way out.

As J.D. says, 'Your own existence is the surest thing you're of. Everything else is problematical. And it has yet to be shown that it's better to be alive. It's entirely guesswork.' And J.D. this is *your* line, 'If you find yourself at sea, in a small unlighted boat, alone in the darkness of a cloudless night, and if you gaze into the darkness of the space between the stars, then keep wide awake, and if your heart is filled with wonder and your mind is filled with peace, there is a chance that you will understand.' It's *your* line.

This body is just a handle through which we get to perceive the world; and it and the world are made of the same stuff. If the body weren't made of the Undivided, it wouldn't stay on the sidewalk, and if it weren't made of the Infinite, it could neither see, hear, taste nor smell.

Thank you! I don't need it any longer. It can go.

That which I had to find, I found.

Thank You!

The life of the Old Man in the shrine is the observational evidence that the Underlying Existence may be addressed as Mother, and that it's possible to reach Her. And there was that Old Boy in China, whose name we don't know, who saw that he was not abandoned by his *Immortal Mother Above*. It was he who said, "In my destitution, She comes and bends over me. To Her only I bow, trusting Her now and forever." And J.D., it was *you* who remembered that line, "*Thou hast lifted all my sorrow with the vision of Thy face, and the magic of Thy Beauty has bewitched my mind.*"

To Her only we go, trusting Her now and forever.

"Be swift my soul to answer Her, be *jubilant* my feet!"

And J.D. please tell Avila not to cry for me.

Yours in That to which the tear drop goes,

Your Little One, Farewell!

And I remembered what was said in the Upanishad, "A man, who knows Him truly, passes over death. *There is no other path to go.*" And I remembered Ruth's song.

> *"Now has come the full turning of the wheel.*
> *Now has come the time to give up the unreal,*
> *For the man who dies before he dies,*
> *Does not die when he dies, - when he dies."*

May Mother protect her! She is out of our hands.

*And may the likes of her attain the fulfillment of their desires!*

# INDEX